Tumor Marker & Carcinogenesis

By
Dr. Manjul Tiwari

Tumor Marker & Carcinogenesis

By

Dr. Manjul Tiwari

MDS, Senior Lecturer (Oral Pathology & Microbiology)
School of Dental Sciences, Sharda University
Greater Noida (Uttar Pradesh), India

River Publishers

Aalborg

ISBN: 978-87-92329-37-0

Published, sold and distributed by:
River Publishers
PO Box 1657
Algade 42
9000 Aalborg
Denmark

Tel.: +4536953197
www.riverpublishers.com

Dedicated to
My Papa, Mummy, Bubu & Bobby

Contents

Acknowledgement   ix

1 Introduction   **1**
   1.1   Diagnostic   2
   1.2   Prognostic   2
   1.3   Therapeutic   3

2 History   **7**

3 Classification   **11**
   3.1   Enzymes   11
   3.2   Tissue Receptors   11
   3.3   Antigens   14
   3.4   Oncogenes   14
   3.5   Hormones   14

4 Mechanism   **15**
   4.1   Stages of Carcinogenesis   15
   4.2   Oncogenes Carcinogene & Tumor Marker   17
   4.3   Activation   19

5 Importance   **23**
   5.1   Doubling Time Of Tumor Cells   25
   5.2   Growth Fraction   26
   5.3   Cell Production and Loss   26
   5.4   Cancer Chemotherapy   26
   5.5   Latnt Period of Tumors   26

6 Tumor Marker & Carcinogenesis   **29**
  6.1 Malignant Tumors   33

7 Management   **41**

8 Tumor Marker in Relation to Carcinogenesis   **49**
  8.1 Tumour Growth Markers   49
  8.2 Markers of Tumour Suppression and Anti-Tumour Response   49
  8.3 Angiogenesis Markers   50
  8.4 Markers of Tumour Invasion and Metastatic Potential   50
  8.5 Cell Surface Markers   50
  8.6 Intracellular Markers   50
  8.7 Markers of Anomalous Keratinisation   50
  8.8 Arachidonic Acid Products   50
  8.9 Oncogenes   50
  8.10 Two–Hit Hypothesis   57
  8.11 p53 (Guardian of Genome)   61
  8.12 Guardian of Genome   62
  8.13 Tumour Growth Markers   89
  8.14 Tumour Suppression and Carcinogenesis   101
  8.15 Markers of Tumour Invasion and Metastatic Potential   105
  8.16 Cell Surface Markers   112
  8.17 Intracellular Markers   113
  8.18 Markers of Anomalous Keratinisation   117
  8.19 Arachidonic Acid Products   122
  8.20 Enzymes   122
  8.21 Molecular Markers of the Risk of Carcinogenesis   122
  8.22 Survivin   123
  8.23 Table Show Survivin Expression   124

9 Discussion   **127**

10 Summary and Conclusion   **131**

Acknowledgement

Foremost in my mind is my Father Dr. Murli Dhar Tiwari, Director, Indian Institute of Information and Technology, Allahabad, my Mother Dr. Iti Tiwari and my Sister Dr. Maneesha Tiwari, about whom it suffices to say—I am fortunate and grateful. For their guidance, supporting, Helping and Guiding me at any time I needed in course of this work.

I would like to extend my heartiest regards and gratefulness to Hon'ble Mr. Pradeep Kumar Gupta, Chairman (Trustee), Hon'able Yatender Kumar Gupta Vice Chairman (Trustee), Hon'ble J. P. Gupta, Vice Chancellor, Prof. S. K. Khanna (PA to Vice Chancellor), Dr. Jagadeesh H (Principal, School of Dental Sciences), Dr. Deepak Bhargava (HOD-Oral Pathology), Sharda University, and also My Guide Dr. N. N. Singh (Prof & HOD, Kothiwal Dental College and Research Centre) India for his enthusiastic, expert guidance, their kind support and providing facilities to complete the present book.

I would really like to extend my heartiest regards and thanks to Mr. Mohit Tiwari (Brother), Mrs Tripti Tiwari (Bhabhi) and My Best Friend Dr. Ruchir Tripathi who supported and helped in every possible way throughout this work and made sure this work is completed as per schedule.

Where would I be without my Wife? I should thank my wife, Manju Pandey for her patience and forbearance whilst I have spent hundreds of hours working on it. Through her dedication to editing and her support in making sure I finally finished this work, it was all possible. Words fail me to express my appreciation to my wife and my family whose dedication, love and persistent confidence in me, has taken the load off my shoulder. I owe her for being unselfishly let her intelligence, passions, and ambitions collide with mine. Therefore, I would also thank My In-Laws's which include Mr. R R Pandey–Father–in-law, Mrs. N D Pandey–Mother-in-law, Mr. Sarvan Pandey and His wife as well as daughter (Gudiya), Mr. Dev Pandey and Dr. Vinay Pandey (Brother-in-laws) for letting me take her hand in marriage, and accepting me as a member of the family, warmly. Furthermore, to My family and My In Laws with their thoughtful support, thank you (Speechless).

I can reassure the reader that as this may be my last direct work on vocabulary as such, at least for a while, I have put a bit of "heart and soul" into it! Therefore, I hope that you will very much enjoy this work as well as find it immensely educative!

There are many others whose names could not be included in this column.That does not mean I am ignoring them; it simply means that they deservemore than my expressions in writing.

Lastly, I vividly acknowledge the one who has made it all possible but perhaps not worthwhile to mention.

Dr. Manjul Tiwari

1

Introduction

Tumor markers are a group of proteins (oncoprotein, immunoglobulin, albumin, globulin), hormones (adrenal corticotropic hormone (ACTH), calcitonin, catecholamines), enzymes (acid phosphatase, alkaline phosphatase, amylase, creatine kinase), receptors (estrogen receptor, progesterone receptor, interleukin-2 receptor, and epidermal growth factor receptor), and other cellular products that are over expressed (produced in higher than normal amounts) by malignant cells. Tumor markers are usually normal cellular constituents that are present at normal or very low levels in the blood of healthy persons.

Detection of a higher-than-normal serum level by radioimmunoassay or immunohistochemical techniques usually indicates the presence of a certain type of cancer. Currently, the main use of tumor markers is to assess a cancer's response to treatment and to check for recurrence. In some types of cancer, tumor marker levels may reflect the extent or stage of the disease and can be useful in predicting how well the disease will respond to treatment. A decrease or return to normal in the level of a tumor marker may indicate that the cancer has responded favorably to therapy. If the tumor marker level rises, it may indicate that the cancer is spreading. Finally, measurements of tumor marker levels may be used after treatment has ended as a part of follow-up care to check for recurrence.

Cancer may be regarded as a group of diseases characterized by an (i) abnormal growth of cells (ii) ability to invade adjacent tissue and even distant organs and (iii) the eventual death of the affected patient if the tumor has progressed beyond that stage when it can be successfully removed. Cancer can occur at any site or tissue of the body and may involve any type of cells.

In 1995 the south East Asia Region of WHO found that there is great majority of cancers of the oral cavity in India. These and other international variations in the pattern of oral cancer are attributed to multiple factors such

Manjul Tiwari (MDS), Tumor Marker & Carcinogenesis, 1–6.

as environmental factors, food habits, life style, genetic factor or even inadequacy in detection and reporting of cases. Oral cancers are also predominantly environment related and have socio–cultural relationship.

Majority of malignancies arising in oral mucosa are epithelial in origin approximately 90% of being squamous cell carcinomas.

Oral Carcinoma is the sixth most common malignancy, and is a major cause of cancer morbidity and mortality worldwide. Globally, about 500.000 new oral and pharyngeal cancers are diagnosed annually, and three quarters of these are from the developing world, including about 65.000 cases from India. Differences have been observed in the clinico-pathological and molecular pathological profile in tobacco-smoking and alcohol-associated oral cancers in the USA, UK, France, Japan, and elsewhere, as well as chewing-tobacco-associated oral cancers, particularly in the Indian subcontinent.

Management of oral carcinoma includes early diagnosis, accurate assessment of prognosis and proper therapeutic intervention.

Tumor markers play an important role in all the aspect of management of oral cancer.

The function of tumor marker can be characterized under the heading of:

1.1 Diagnostic

The growing neoplasm consists of a cancer cell cluster, a cancer-supporting stroma and cancer vessels which produces antigen, and cancer vessel-related substances, respectively. Therefore, these three substances are important in early cancer detection. Furthermore, because they are identical to tumor-specific tumor markers, tumor-associated tumor markers, and growth-related tumor markers, they play an important role in the tumor marker diagnosis of oral cancer.

Tumor markers have been identified in several types of cancer, including malignant melanoma, multiple myeloma, and bone, breast, colon, gastric, liver, lung, ovarian, pancreatic, prostate, renal, uterine cancers as well as oral cancers. Serial measurements of a tumor marker are often an effective means to monitor the course of therapy. Some tumor markers can provide us with information used in staging cancers.

1.2 Prognostic

Some types of cancer typically grow and spread faster than other types. But even within a cancer type some cancers will grow and spread more rapidly or may be more or less responsive to certain treatments. Some newer tumor markers help show how aggressive a particular cancer is, or even how well it might respond to a particular drug.

The majority of tumor markers are used to monitor patients for recurrence of tumors following treatment. In addition, some markers are associated with a more aggressive course and higher relapse rate of tumor and have value in prognosis of the oral cancer as well as some help predict the response to treatment.

1.3 Therapeutic

The most important use for tumor markers is to monitor patients being treated for cancer, especially advanced cancer. If a tumor marker is available for a specific type of cancer, it is much easier to measure it to see if the treatment is working rather than to repeated x-rays, computed tomography (CT) scans, bone scans, or other complicated tests. It is also less expensive.

If the marker level in the blood goes down, it is almost always a sign that the treatment is having an effect. On the other hand, if the marker level goes up, then the treatment probably should be changed. (One exception is if the cancer is very sensitive to a particular chemotherapy treatment. In this case, the chemotherapy can cause many cancer cells to rapidly die and release large amounts of the marker, which will cause the level of the marker in the blood to temporarily rise.)

In clinical medicine their utility has been as laboratory test to support the diagnosis but some are also of value in determing the response to therapy and in indicating relapse during the follow up period.

Carcinogenesis (meaning literally, the creation of cancer) is the process by which normal cells are transformed in to cancer cells.

Carcinogenesis is a multistep process resulting from the sequential perturbation of both positive and regulatory networks that normally allow the somatic cell to live a cooperative existence within the society of normal cells that comprise an organism. Normal cells even programmed to give their own life for the good of the organism. Any genetic or epigenetic changes that allow a cell to escape these societal constraints represent a step toward cancer. Survival of the fittest cells allows for the clonal expansion of progeny cells with ever increasing numbers of genetic or epigenetic changes that favor even greater antisocial and selfish behavior of the cancer cell within the organism.

Oral cancer is the most common cancer and constitutes a major health problem in developing countries, representing the leading cause of death. It could be justifiably argued that the proliferation of literature on the molecular basis of cancer has out patched the growth of even the most malignant tumors. Carcinogenesis is a multistep process in which genetic events lead to the disruption of the normal regulatory pathways that control basic cellular functions including cell division, differentiation, and cell death.

Cell division is physiological process that occurs in almost all tissues under many circumstances. Normally homeostasis means the balance between proliferation and programmed cell death, usually in the form of apoptosis and is maintained tightly by regulating both the processes to ensure the integrity of organs and tissues. Mutations in DNA that lead to cancer disrupt these orderly processes by disrupting the programming that regulates the processes.

Carcinogenesis is a multistep process at both the phenotypic and genetic level. A malignant neoplasm has several phenotypic attributes, such as excessive growth, local invasiveness, and the ability to form distant metastasis. These characteristics are acquired in a stepwise fashion, a phenomenon called tumor progression.

At the molecular level, progression results from accumulation of genetic lesions that in some instances are favored by defects in DNA repair.

Carcinogenesis is caused by mutation of the genetic material of normal cells, which upsets the normal balance between proliferation and cell death. This results in uncontrolled cell division and tumor formation. The uncontrolled and often rapid proliferation of cells can lead to benign tumors. Some types of these may turn in to malignant tumors.

Benign tumors do not spread to other parts of the body or invade other tissues, and they are rarely a threat to life unless they compress vital structures or are physiologically active.

Malignant tumors can invade other organs, spread to distant locations, and become life threatening.

More than one mutation is necessary for carcinogenesis. Only mutations in those certain types of genes which play vital role in cell division, cell death, and DNA repair will cause a cell to loose control of its proliferation.

Among the molecular mechanisms involved in the carcinogenesis, defects in the regulation of programmed cell death (apoptosis) may contribute to the pathogenesis and progression of cancer. Dysregulation of oncogenes and tumor suppressor genes involved in apoptosis are also associated with tumor development and progression.

Genes that regulate apoptosis may be dominant, as are proto-oncogene, or they may behave as cancer suppressor genes.

In addition to the three classes of genes mentioned earlier, a fourth category of genes, those that regulate repair of damaged DNA are also pertinent in carcinogenesis.

DNA repair genes affect cell proliferation or survival indirectly by influencing the ability of the organism to repair nonlethal damage in other genes, including proto-oncogenes, tumor suppressor genes, and genes that regulate apoptosis.

The unregulated growth that characterizes cancer is caused by damage to DNA, resulting in mutations to gene that encode for proteins controlling cell division. Many mutation events may be required to transform a normal cell in to a malignant cell. These mutations can be caused by chemicals or physical agents called carcinogens, by close exposure to radioactive materials or by certain viruses that can insert their DNA in to the human genome.

Mutations occur spontaneously, and may be passed down from one generation to the next as a result of mutations within germ line.

The genetic hypothesis of cancer implies that a tumor mass results from the clonal expansion of a single progenitor cell that has incurred the genetic damage. Clonality of tumors is assessed quite readily in women who are heterozygous for polymorphic X-linked markers, such as the enzyme glucose-6-phosphate dehydrogenase (G6PD) or X-linked restriction fragment length polymorphism.

A malignant neoplasm has several phenotypic attributes, such as excessive growth, local invasiveness, and the ability to form distant metastases. These characteristics are acquired in a stepwise fashion, a phenomenon called tumor progression.

Many forms of cancer are associated with exposure to environmental factors such as tobacco smoke, radiation, alcohol and certain viruses.

A variety of agents increase the frequency with which cells are converted to the transformed condition, they are said to be carcinogenic agents. Carcinogens may cause epigenetic changes or may act directly or indirectly to change the genotype of the cells.

Although tobacco is clearly of major aetiological significance (IARC 1984) the failure of overtly malignant lesions to develop in all tobacco users and the development of oral cancer in all tobacco users and the development of oral cancer in persons with no history of tobacco use suggests that the genesis of oral cancer may also involve other unidentified environmental and host factors.

So experimental carcinogenesis provide an overview of

1. Research based on studies of human material designed to elucidate biological changes in the host tissues which may be pathognomonic of malignant change which includes-

    1) studies on cell proliferation
    2) studies on cell surface changes
    3) biochemical changes in carcinogenesis
    4) changes in hydrolytic enzymes
    5) Role of viruses in oral carcinogenesis

2. Experimental animal studies designed to evaluate the ability of agents to initiate cancer and the biological effects of known and possible carcinogenic agents on oral mucosa.
   Some of the unexpected carcinogens include alcoholic beverages, asbestos, benzene, viruses, radiation and drugs which can be best understand by the studies of-

    1) experimental animal models
    2) factors modifying the response to carcinogens
    3) influence of immune status on experimental carcinogenesis
    4) the role of viruses in experimental tumors

2

History

The interesting salient features in relation to historical aspect of oral carcinoma and tumor markers are as follows:

> Celsus has translated carcinos in to the Latin cancer, also meaning crab. Galen used 'oncos' to describe all tumors, the root for the modern world oncology.

Hippocates described several kinds of cancers. He called benign tumors oncos, Greek for swelling, and malignant tumors carcinos, Greek for crab or crayfish. This name probably comes from the cut surface of a malignant tumor, with a roundest hard center surrounded by pointy projections, vaguely resembling the shape of a crab. He later added the suffix–oma, Greek for swelling, giving the name 'Carcinoma'.

The first modern tumor marker used to detect cancer was human chorionic gonadotropin (HCG), the substance doctors look for in pregnancy tests. Women whose pregnancy has ended but whose uterus continues to be enlarged are tested for the presence of HCG. A high level of HCG in the blood may indicate the presence of a cancer of the placenta called gestational trophoblastic disease (GTD). This cancer continues to produce HCG. Some testicular and ovarian cancers resemble GTD because they arise from reproductive cells called germ cells. These cancers also make HCG, so this marker is used to help in their diagnosis and to monitor their response to therapy.

The hope in the search for tumor markers was that all cancers could someday be detected by a single blood test. Both GTD and germ cell tumors of the ovaries and testicles are too rare to look for these cancers by testing everyone. But cancers such as colon, breast, and lung are more common. A simple blood test that would be able to detect these cancers in their earliest stages could prevent the deaths of millions of people. Many scientists began working toward this goal.

Manjul Tiwari (MDS), Tumor Marker & Carcinogenesis, 7–10.

The first success in developing a blood test for a common cancer was in 1965, when carcinoembryonic antigen (CEA) was found in the blood of some patients with colon cancer. By the end of the 1970s several other blood tests had been developed for different cancers. The new markers were often given numeric labels. There was CA 19-9 for colorectal and pancreatic cancer, CA15-3 for breast cancer, and CA 125 for ovarian cancer. Many others were also discovered, but because they did not show an advantage over the already discovered markers, they were not studied further.

The only tumor marker that currently allows doctors to detect early disease and is used in screening is the prostate-specific antigen (PSA) test. It was discovered around the same time as the others, but it's been in widespread use for screening since the early 1990s because it has some advantages over them. First, it is made only by prostate cells, so a rise in PSA is fairly specific to a prostate problem. And the PSA level usually rises even in early cancers, so most prostate cancers can be detected at an early stage, when they are most likely to be curable. Some men may have an elevated PSA because of other prostate conditions (or prostate cancer that would never need treatment), and some men with prostate cancer may not have an elevated PSA.

The first breast cancer susceptibility gene, *BRCA1*, was identified and cloned in 1994 by Miki and colleagues after an intensive search (Miki *et al.* 1994). A year later, a second breast cancer susceptibility gene, BRCA2 was identified (Wooster *et al.* 1995, Tavtigian *et al.* 1996).

Like BRCA1, BRCA2 probably regulates the activity of other genes and plays a critical role in embryo development. The BRCA2 gene was discovered in 1995 by Professor Michael Stratton and Dr Richard Wooster (Institute of Cancer Research, UK).

Dr. Robert Bast and his research team first isolated the monoclonal antibody in 1981. CA-125, also known as CA125, is an abbreviation for cancer antigen 125. CA-125 is a tumor marker or biomarker that may be elevated in the blood of some people with specific types of cancers. CA-125 is a mucinous glycoprotein and the product of the MUC16 gene. It is best known as a marker for ovarian cancer, but it may also be elevated in other malignant cancers, including those originating in the endometrium, fallopian tubes, lungs, breast and gastrointestinal tract. CA-125 was initially detected using the monoclonal antibody designated OC125.

Lab Corp, a large US clinical laboratory testing company, began offering AFP screening tests in the early 1980's. Alphafetoprotein (AFP) is a protein produced in the developing embryo and fetus. In humans, AFP levels decrease gradually after birth, reaching adult levels by 8 to 12 months.

Normal adult AFP levels are low, but detectable; however, AFP has no known function in normal adults.

p53 is a nucleo phosphoprotein which act as tumor suppressor comprising 393amino acids, and was discovered in 1970. It is located on chromosome 17p13.1 and is the single most common target for genetic alteration in oral tumors. The physiologic function of the p53 protein is that of preventing accumulation of genetic damage in cells either by allowing for repair of the damage before cell division or by causing death of cell.

The first oncogene was discovered in 1970 and was termed SRC (pronounced *SARK*). Src was in fact first discovered as an oncogene in a chicken retrovirus. Experiments performed by Dr G. Steve Martin of the University of California, Berkeley demonstrated that the SRC was indeed the oncogene of the virus.

In 1976 Drs. J. Michael Bishop and Harold E. Varmus of the University of California, San Francisco demonstrated that oncogenes were defective proto-oncogenes, found in many organisms including humans. For this discovery Bishop and Varmus were awarded the Nobel Prize in 1989.

Myc (cMyc) is a protooncogene, which is over expressed in a wide range of human cancers. When it is specifically-mutated, it loses its function and becomes an oncogene. Myc gene was first discovered in Burkett's lymphoma patients. In Burkett's lymphoma, cancer cells show chromosomal translocations, in which Chromosome 8 is frequently involved. Cloning the break point of the fusion chromosomes revealed a gene that was similar to myelocytomatosis viral oncogene (v-Myc). Thus, the new found cellular gene was named c-Myc.

Many other tumor markers have been found in recent years and are currently under study. Some of these are different from traditional markers, which were proteins found in the blood.

Table 2.1 Historical Advent in Tumor Markers

Name of Tumor Marker	Year of Discovery
Carcinoembryonic antigen (CEA)	1965
CA 19-9, CA15-3, CA 125	1970s
p53	1970
SRC	1970
Defective proto-oncogenes,	1976
Alphafetoprotein (AFP)	1980's
Prostate-specific antigen (PSA)	Early 1990s
BRCA1	1994
BRCA2	1995

Early theories of carcinogenesis includes-

1. Surfeit of black bile
2. Omnis cellula ex cellula
3. Irritation hypothesis
4. Embryonic hypothesis
5. Parasitic hypothesis

Recent theories of carcinogenesis includes-

1. Chemical carcinogenesis- derived observations by major line of mechanistic oncology
2. viral theory of carcinogenesis
3. two stage mechanism of carcinogenesis- two processes- initiation, promotion followed by progression.

According to the old theory of irritation as the cause of cancer, neoplasia was a kind of extension of hyperplasia. The irritational theory was related to the traumatic origin of tumors. virchew supported the essential role of gross mechanical insult in the origin of tumors. the embryonic hypothesis suggested that tumors originated from undifferentiated embryonal cells persist in many adult tissues. the embryonic character of these cells endowed them with a "marked capacity for proliferation".

The dominant hypothesis today is that neoplastic development is a multistage progressive process involving multiple genetic changes.

Cancer develops through four definable stages: initiation, promotion, progression and malignant conversion. These stages may progress over many years. The first stage, initiation, involves a change in the genetic makeup of a cell. During promotion, the mutated cell is stimulated to grow and divide faster and becomes a population of cells. During progression, there is further growth and expansion of the tumor cells over normal cells.

Early chemical carcinogenesis experiments were performed in the beginning of the 20th century.

More recently, an epidemiological study by Maier *et al.* (1990) showed that 90% of all patients with head and neck cancer consumed alcohol regularly in quantities twice the amount of a control group with a significant dose–response relationship.

3

Classification

According to various sources tumor markers can be classified into following subtypes:

1. According to Tumor Marker Tests. http:// July, 2002 [cited April 4, 2003]. Resources

Tumor marker can be classified as five basic types:

1. Enzymes
2. Tissue receptors
3. Antigens
4. Oncogenes
5. Hormones

3.1 Enzymes

Many enzymes that occur in certain tissues are found in blood plasma at higher levels when the cancer involves that tissue. Enzymes are usually measured by determining the rate at which they convert a substrate to an end product, while most tumor markers of other types are measured by a test called an immunoassay. Some examples of enzymes whose levels rise in cases of malignant diseases are acid phosphatase, alkaline phosphatase, amylase, creatine kinase, gamma glutamyl transferase, lactate dehydrogenase, and terminal deoxynucleotidyl transferase.

3.2 Tissue Receptors

Tissue receptors, which are proteins associated with the cell membrane, are another type of tumor marker. These substances bind to hormones and growth factors, and therefore affect the rate of tumor growth. Some tissue receptors must be measured in tissue samples removed for a biopsy, while

Manjul Tiwari (MDS), Tumor Marker & Carcinogenesis, 11–14.
© *2012 River Publishers. All rights reserved.*

Table 3.1 Oncogenes can be further classified as:

Selected oncogenes, their mode of activation and associated human tumors

Category	Protooncogene	Mode of Activation		Associated Human Tumor
Growth factors				
PDGF – β chain	SIS	Overexpression		Astrocytoma
				Osteosarcoma
Fibroblast growth factors	HST–1	Overexpression		Stomach cancer
	INT–2	Amplification		Bladder cancer
				Breast cancer
				Melanoma
TGF α	TGF α	Overexpression		Astrocytoma
				Hepatocellular carcinomas
HGF	HGF	Overexpression		Thyroid cancer
Growth factor receptor				
EGF – receptor family	ERB–B1 (ECFR)	Overexpression		Squamouung, gliomas
	ERB–B2	Amplification		Breast and ovarian cancer
CSF – 1 receptor	FMS	Point mutation		Leukemia
Receptor for neurotrophic factors	RET	Point mutation		Multiple endocrine neoplasia 2A and B, familial medullary thyroid carcinomas
PDGF receptor	PDGF–R	Overexpression		Gliomas
Receptor for stem cell (steel) factor	KIT	Point mutation		Gastrointestinal stromal tumors and other soft tissue tumors
Proteins involved in signal transduction				
GTP-binding	K–RAS	Point mutation		Colon, lung and pancreatic tumors
	H–RAS	Point mutation		Bladder and kidney tumors
	N–RAS	Point mutation		Melanomas, hematologic malignancies
Nonreceptor tyrosine kinase	ABL	Translocation		Chronic myeloid leukemia
				Acute lymphoblastic leukemia
RAS signal transduction	BRAF	Point mutation		melanomas
WNT signal transduction	B – catenin	Point mutation		Hepatoblastomas, Hepatocellular carcinoma
		Overexpression		
Nuclear regulatory proteins				
Transcriptional activators	C–MYC	Translocation		Burkitt lymphoma
	N–MYC	Amplification		Neuroblastoma, small cell carcinoma of lung
	L–MYC	Amplification		
Cell – Cycle regulators				
Cyclins	CYCLIN – D	Translocation		Mantle cell lymphoma
	CYCLIN – E	Amplification		Breast and esophageal cancers
		Overexpression		Breast cancers
Cyclin – dependent kinase	CDK4	Amplification or point mutation		Gliblastoma, melanoma, sarcoma

Table 3.2 Selected Tumor Suppressor Genes in Human Neoplasms are:

Subcellular Location	Gene	Function	Tumors Associated with Somatic Mutations	Tumors Associated with Inherited Mutations
Cell surface	TGF – β Receptor E-cadherin	Growth inhibition Cell adhesion	Carcinomas of colon, carcinomas of stomach	Unknown, familial gastric cancer
Inner aspect of plasma membrane	NF-1	Inhibition of RAS signal transduction and of p21 cell cycle inhibitor	Neuroblastomas	Neurofibromatosis type 1 and sarcomas
Cytoskeleton	NF-2	Cytoskeletal stability	Schwannomas and meningiomas	Neurofibromatosis type 2 acoustic schwannomas and meningiomas
Cytosol	APC/β –catein PTEN SMAD 2 AND SMAD 4	Inhibition of signal transduction PI-3 kinase signal transduction TGF-B signal transduction	Carcinomas of stomach, colon, pancreas, melanomas Endometrial and prostate cancers Colon, pancreas tumors	Familial adenomatous polyposis coli/ colon cancer unknown
Nucleus	RB P53 Wt-1 P16(inka) BRCA -1AND BRCA 2 KLF6	Regulation of cell cycle Cell cycle arrest and apoptosis in response to DNA damage Nuclear transcription Regulation of cell cycle by inhibition of cyclin dependent kinases DNA repair Transcription factor	Retinoblastoma; osteosarcomas, carcinoma of breast, colon, lung most human cancers wilms tumor pancreatic, breast and esophageal cancers unknown prostate	Retinoblastoma; Osteosarcomas li-fraumeni syndrome, multiple carcinomas and sarcomas wilm tumors malignant melanomas carcinoma of female breast and ovary, carcinomas of male breast unknown

others are secreted into the extra cellular fluid (fluid outside the cells) and may be measured in the blood. Some important receptor tumor markers are estrogen receptor, progesterone receptor, interleukin-2 receptor, and epidermal growth factor receptor.

3.3 Antigens

Oncofetal antigens are proteins made by genes that are very active during fetal development but function at a very low level after birth. The genes become activated when a malignant tumor arises and produce large amounts of protein. Antigens comprise the largest class of tumor marker and include the tumor-associated glycoprotein antigens. Important tumor markers in this class are alpha-fetoprotein (AFP), carcinoembryonic antigen (CEA), prostate specific antigen (PSA), cathespin-D, HER-2/neu, CA-125, CA-19-9, CA-15-3, nuclear matrix protein, and bladder tumor-associated antigen.

3.4 Oncogenes

Some tumor markers are the product of oncogenes, which are genes that are active in fetal development and trigger the growth of tumors when they are activated in mature cells. Some important oncogenes are BRAC-1(located on chromosome 17q12-21 is recently discovered tumor suppressor genes that are associated with the occurrence of breast and several other cancers), myc, p53, RB (retinoblastoma) gene (RB), and Ph (Philadelphia chromosome).

3.5 Hormones

The fifth type of tumor marker consists of hormones. This group includes hormones that are normally secreted by the tissue in which the malignancy arises as well as those produced by tissues that do not normally make the hormone (ectopic production). Some hormones associated with malignancy are adrenal corticotropic hormone (ACTH), calcitonin, catecholamines, gastrin, human chorionic gonadotropin (hCG), and prolactin.

ACCORDING TO KERN SE "Progressive genetic abnormalities in human neoplasia" Robins and Cotran "Pathologic basis of disease" 7TH edition

Association of various tumor markers with human tumor.

4

Mechanism

Cancer is ultimately a disease of genes. In order for cells to start dividing uncontrollably, genes which regulate cell growth must be damaged.

Protooncogenes are genes which promote cell growth and mitosis, a process of cell division, and tumor supressor genes discourage cell growth, or temporarily halts cell division from occuring in order to carry out DNA repair. Typically a series of several mutations to these genes are required before a normal cell transforms in to a cancer cell.

Oral carcinogenesis is a multistep process in which genetic events lead to the disruption of the normal regulatory pathways that control basic cellular functions including cell division, differentiation, and cell death.

4.1 Stages of Carcinogenesis

Cancer develops through four definable stages: initiation, promotion, progression and malignant conversion. These stages may progress over many years. The first stage, initiation, involves a change in the genetic makeup of a cell. This may occur randomly or when a carcinogen interacts with DNA causing damage. This initial damage rarely results in cancer because the cell has in place many mechanisms to repair damaged DNA. However, if repair does not occur and the damage to DNA is in the location of a gene that regulates cell growth and proliferation, DNA repair, or a function of the immune system, then the cell is more prone to becoming cancerous.

During promotion, the mutated cell is stimulated to grow and divide faster and becomes a population of cells. Eventually a benign tumor becomes evident. In human cancers, hormones, cigarette smoke, or bile acids are substances that are involved in promotion. This stage is usually reversible as evidenced by the fact that lung damage can often be reversed after smoking stops.

Manjul Tiwari (MDS), Tumor Marker & Carcinogenesis, 15–21.
© 2012 *River Publishers. All rights reserved.*

The progression phase is less well understood. During progression, there is further growth and expansion of the tumor cells over normal cells. The genetic material of the tumor is more fragile and prone to additional mutations. These mutations occur in genes that regulate growth and cell function such as oncogenes, tumor suppressor genes, and DNA mismatch-repair genes. These changes contribute to tumor growth until conversion occurs, when the growing tumor becomes malignant and possibly metastatic. Many of these genetic changes have been identified in the development of colon cancer and thus it has become a model for studying multi-stage carcinogenesis.

Human papillomaviruses have evolved proteins to control the growth of the epithelial cells they infect. This was a necessity since these viruses require a metabolically active, dividing cell in order to complete their life cycle. In particular, the E6 and E7 proteins have the ability to abrogate growth and differentiation controls that would otherwise prevent epithelial cell growth and stymie viral propagation. The "E" designation indicates an early gene, meaning a viral gene that is turned on early in the process of infecting a cell.

The HPV genome typically consists of nine open-reading frame sequences, located on only one of the strands of DNA, and is divided into seven early-phase genes (E) and two late-phase genes (L). The early genes serve to regulate the transcription of DNA, while the late genes encode for proteins involved in viral spread, such as capsid proteins. The E1 and E2 gene products are more specifically involved in regulating the transcription and replication of viral proteins. These different gene regions and gene products provide the basis on which molecular detection methods have been created.

The human papillomavirus genome. The "E" designation indicates an early viral protein which is expressed early in a vegetative infection. Similarly, the "L" designation indicates a late viral gene, usually involved in viral protein coats.

The E7 protein targets the retinoblastoma protein, a critical component of cell cycle control. The retinoblastoma protein (Rb) in the unphosphorylated state binds to and sequesters transcription factors necessary for progression through the cell cycle, particularly E2F and related proteins. This prevents cells from dividing until E2F becomes available in the unbound state, usually by release from Rb. In normal cellular physiology, this release is accomplished by Rb phosphorylation by one of the cyclin-dependent kinases. In the case of a papillomavirus infection, E2F release is due to binding of Rb by viral E7 protein.

E7 Effects on Rb. E7 binding of RB leads to release of sequestered E2F, enabling the cell cycle to progress.

4.2 Oncogenes Carcinogene & Tumor Marker

Normal cell proliferation is controlled by growth factors and cytokines that act on the cell membrane by triggering the cascade of biochemical signals (a process called signal transduction). These signals control the genes that regulate cell growth and division. Oncogenes are altered forms of normal cellular genes called proto-oncogenes that are involved in this cascade of events. They may mutate spontaneously through interaction with viruses, chemicals, or by physical means.

These cellular genes were first discovered by the Noble laureate Michael Bishop and Harold Varmus as passengers within the genome of acute transforming retroviruses, which cause rapid induction of tumors in animals and can also transform animal cells in vitro. Molecular dissection of their genomes revealed the presence of unique transforming sequences not found in the genomes of nontransforming retroviruses. Most surprisingly, molecular hybridization revealed that the viral oncogenes were almost identical to sequences found in the normal cellular DNA.

Most surprisingly, molecular hybridization revealed that the v-onc (viral oncogenes) sequences were almost identical to sequences found in the normal cellular DNA. From this evolved the concept that during evolution, retroviral oncogenes were transuduced (captured) by the virus through a chance recombination with the DNA of a (normal) host cell that had been infected by the virus. Because they were discovered initially as viral genes, proto-oncogenes are named after their viral homologs. Each v-oncogene is designated by the oncogene to the virus from which it was isolated. Thus the v-onc contained in feline sarcoma virus is referred to as v-fes, whereas the oncogene in simian sarcoma virus is called v-sis.

V-oncs are not present in several cancer causing RNA viruses. Example is a group of so called slow transforming viruses that cause leukemia's in rodents after a long latent period. The mechanism by which they cause neoplastic transformation implicates proto-oncogenes. Molecular dissection of the cells transformed by these leukemia viruses has revealed that the proviral DNA is always found to be integrated.

(Inserted) near a proto-oncogene. One consequence of proviral insertion near a proto-oncogene is to induce a structural change in the cellular gene, thus converting it in to a cellular oncogene (c-onc). The strong retroviral promoters inserted in the proto-oncogenes lead to dysregulated expression of the cellular gene. This mode of proto-oncogene activation is called insertional mutagenesis.

Oncogenes are altered growth promoting regulatory genes that governs the cells signal transduction pathways and mutations of these genes leads to either overproduction or increased function of the excitatory proteins. Although oncogenes alone are not sufficient to transform epithelial cells, they appear to be important initiators of the process, and are known to cause cellular changes through mutation of only one gene copy.

Oncogenes or cancer-causing genes are derived from proto-oncogenes. Proto-oncogenes are cellular genes that promote normal growth and differentiation. Proto-oncogenes may become oncogenic by retroviral transduction or by influences that their behaviour in situ, thereby converting them in to cellular Oncogenes.

Several Oncogenes have been implicated in oral carcinogenesis. Aberrant expression of the proto-oncogene epidermal growth factor receptor (EGFR/c-erb 1), members of the ras gene family, c-myc, int-2, hst-1, PRAD-1, and bcl-1 is believed to contribute towards cancer development. Molecular hybridization of oral carcinogens revealed v-onc (viral oncogene).

Deregulation of growth factors occurs during oral carcinogenesis through increased production and autocrine stimulation. Aberrant expression of transforming growth factor $\acute{\alpha}$ (TGF-$\acute{\alpha}$) is reported to occur early in oral carcinogenesis, first in hyperplasic proliferation by EGFR in an autocrine and paracrine fashion. TGF-$\acute{\alpha}$ is believed to stimulate angiogenesis and has been reported to be found in normal mucosa in patients who subsequently develop a second primary carcinoma.

Usually oncogenes are dominant as they contain gain-of-function mutations, while mutated tumor suppressors are recessive as they contain loss-of-function mutations. Each cell has two copies of a same gene, one from each parent, and under most cases gain of function mutation in one copy of a particular proto-oncogene is enough to make that gene a true oncogene, while usually loss of function mutation need to happen in both copies of a tumor suppressor gene to render that gene completely non-functional. However, cases exist in which one loss of function copy of a tumor suppressor gene can render the other copy non-functional, and this is called the *dominant negative effect*. This is observed in many p53 mutations.

Genes are the means by which a cell produces proteins, each of which has a very specific role. A mutated gene can cause overproduction, underproduction, or alteration of a protein that may be unable to carry out its purpose.

Oncogenes typically produce more of their protein product when mutated, while tumor suppressor genes typically produce less of their protein product when mutated.

Rb is a classic example of a tumor suppressor gene. Tumor suppressor genes are normal genes whose loss of function predisposes the cell to cancer. These genes are involved in various important processes critical to cellular homeostasis, including differentiation, DNA repair, control of the cell cycle and apoptosis. Complete loss of function of the retinoblastoma gene is seen in conjunction with a number of tumors including retinoblastomas and melanomas. The relative carcinogenic potency of the various human papillomavirus types is partially explained by the relative avidity of their respective E7 proteins for Rb, *e.g.*, the more tightly a given viral E7 binds Rb, the greater the oncogenic potential of the virus.

When a proto-oncogene is altered to become an oncogene, the pathway of cell growth and proliferation become altered. This may lead to the abnormal growth of cells (neoplastic transformation). More than 100 oncogenes have been identified. An example of an oncogene is the K-ras gene that is mutated in colon cancer cells.

4.3 Activation

This is brought about by two broad categories of changes-

- Changes in the structure of the gene, resulting in the synthesis of an abnormal gene product (oncoprotein) having an aberrant function.
- Changes in the regulation of gene expression, resulting in enhanced or inappropriate production of the structurally normal growth promoting protein.

Activation is done by-

    a. Point mutation
    b. Chromosomal rearrangements
    c. Gene amplification

4.3.1 Point mutations

The ras oncogenes represent the best example of activation by point mutations. A large number of human tumors carry ras mutations. The frequency of such mutations varies with different tumors, but in some types it is high. A single nucleotide base may be substituted by a different base, resulting in a point mutation. Point mutations reduce the GTPase activity of the ras proteins.

Point mutations (single base changes) can lead to over activity or inactivity of gene products. These are common in genes such as K-ras and p53. Amplifications and rearrangements have frequently been reported in malignant neoplasms, with both amplification and rearrangement affecting excitatory pathway genes, whereas rearrangement can also inactivate inhibitory pathway genes.

In general, carcinomas have mutations of K-ras, whereas hematopoietic tumors bear N-ras mutations. ras mutations are infrequent or even nonexistent in certain other tumors. Therefore although ras mutations are extremely common, their presence is not essential for carcinogenesis. In addition to ras, activating point mutations have been found in the c-fms gene in some cases of acute myeloid leukemia.

4.3.2 Chromosomal rearrangements

Two types of chromosomal rearrangements can activate proto-oncogenes-translocations and inversions. Of these chromosomal translocations are much more common. Translocations can activate proto-oncogenes in two ways-

1. In lymphoid tumors, specific translocations result in over expression of proto-oncogenes by placing them under the regulatory elements of the immunoglobulin or T-cell receptor loci.
2. In many hematopoietic tumors, the translocation allows normally unrelated sequences from two different chromosomes to recombine and form hybrid genes that encode growth promoting chimeric proteins.

Translocation-induced over-expression of a proto-oncogene is best exemplified by Burkitt lymphoma. In Burkitt lymphoma, the most common form of translocation, results in the movement of the c-myc containing segment of chromosome 8 to chromosome 14q band 32.

The molecular mechanisms of the translocation- associated activation of c-myc are variable, as are the precise breakpoints within the gene. In some cases the translocation renders the c-myc gene subject to relentless stimulation by the adjacent enhancer element of the immunoglobulin gene. In others, the translocation causes mutations in the regulatory sequences of the myc gene. In all instances, the coding sequences of the gene remain intact, and the myc gene is constitutively expressed at high levels.

The Philadelphia chromosome characteristic of chronic myeloid leukemia and a subset of acute lymphoblastic leukemias, provides the prototypic example of an oncogene formed by fusion of two separate genes. In these cases, a reciprocal translocation between chromosome 9 and 22 relocates a truncated portion of the proto-oncogene c-abl (from chromosome 9) to the bcr gene encodes for a chimeric protein that has tyrosine kinase activity. Although the translocations are cytogenetically identical in chronic myeloid leukemia and acute lymphoblastic leukemias, they differ at the molecular level. In chronic myeloid leukemia the chimeric protein has a molecular weight of 210KD, whereas in the more aggressive acute leukemias, a slightly different, 180KD, abl-bcr fusion protein is formed.

4.3.3 Gene amplification

Activation of proto-oncogenes associated with over expression of their products may result from reduplication and manifold amplification of their DNA sequences. Such amplification may produce several hundred copies of the proto-oncogene in the tumor cell. The amplified genes can be readily detected by molecular hybridization with the DNA probe. Two mutually exclusive patterns are seen: multiple small, chromosome like structures called double minutes (dms) or homogenous staining regions (HSRs) derived from the family of amplified genes in to new chromosomes. Gene amplification are involved in breast, ovarian, and lung carcinomas. Gene amplification plays a major role during development when it is responsible for the programmed increase of gene expression.

The most interesting cases of amplification involve N-myc in neuroblastomas and c-erbB2 in breast cancers and this amplification is associated with poor prognosis.

Other genes frequently amplified include c-myc (breast, ovarian and lung carcinomas) and cyclin D (breast carcinomas and squamous cell carcinomas).

5

Importance

Currently, there are over 60 analyses that are used as tumor markers. All of the enzymes and hormones mentioned above have been approved as tumor markers by the Food and Drug Administration (FDA), but most of the others are not; they have been designated for investigation purposes only. The following list describes the most commonly used tumor markers approved by the FDA for screening, diagnosis, or monitoring of cancer.

1. Alpha-fetoprotein (AFP)
2. CA-125
3. Carcinoembryonic antigen (CEA)
4. Prostate specific antigen (PSA)
5. Estrogen receptor (ER)
6. Progesterone receptor (PR)
7. Human chorionic gonadotropin (hCG)
8. Nuclear matrix protein (NMP22) and bladder tumor-associated analytes (BTA)

Alpha-fetoprotein (AFP): AFP is a glycoprotein produced by the developing fetus, but its blood levels decline after birth. Healthy adults who are not pregnant rarely have detectable levels of AFP in their blood. The AFP test is primarily used for prenatal diagnoses of spina bifida and other abnormalities associated with cerebrospinal fluid leakage during embryonic development. In adult males and nonpregnant females, an AFP above 300 mg/L is often associated with cancer, although levels in this range may be seen in non malignant liver diseases. Levels above 1000 mg/L are almost always associated with cancer. AFP has been approved by the FDA for the diagnosis and monitoring of patients with non-seminoma testicular cancer. It is elevated in almost all yolk sac tumors and 80% of malignant liver tumors.

CA-125: Measurement of this tumor marker is FDA-approved for the diagnosis and monitoring of women with ovarian cancer. Approximately 75%

Manjul Tiwari (MDS), Tumor Marker & Carcinogenesis, 23–27.

of persons with ovarian cancer shed CA-125 into the blood and have elevated serum levels. Elevated levels of CA-125 are also found in approximately 20% of persons with pancreatic cancer. Other cancers detected by this marker include malignancies of the liver, colon, breast, lung, and digestive tract. Test results, however, are affected by pregnancy and menstruation. Benign diseases detected by the test include endometriosis, ovarian cysts, fibroids, inflammatory bowel disease, cirrhosis, peritonitis, and pancreatitis. CA-125 levels correlate with tumor mass; consequently, this test is used to determine whether recurrence of the cancer has occurred following chemotherapy. Some patients, however, have a recurrence of their cancer without a corresponding increase in the level of CA-125.

Carcinoembryonic antigen (CEA): CEA is a glycoprotein that is part of the normal cell membrane. It is shed into blood serum and reaches very high levels in colorectal cancer. Over 50% of persons with breast, colon, lung, gastric, ovarian, pancreatic, and uterine cancer have elevated levels of CEA. CEA levels in plasma are monitored in patients with tumors that secrete this antigen to determine if second-look surgery should be performed. CEA levels may also be elevated in inflammatory bowel disease (IBD), pancreatitis, and liver disease. Heavy smokers and about 5% of healthy persons have elevated plasma levels of CEA.

Prostate specific antigen (PSA): PSA is a small glycoprotein with protease activity that is specific for prostate tissue. The antigen is present in low levels in all adult males, which means that an elevated level may require additional testing to confirm that cancer is the cause. High levels are seen in prostate cancer, benign prostatic hypertrophy, and inflammation of the prostate. PSA is approved as a screening test for prostatic carcinoma. PSA has been found to be elevated in more than 60% of persons with Stage A and more than 70% with Stage B cancer of the prostate. It has replaced the use of prostatic acid phosphatase for prostate cancer screening because it is far more sensitive. Most PSA is bound to antitrypsins in plasma but some PSA circulates unbound to protein (free PSA). Persons with a borderline total PSA (between 4–10 mg/L), but who have a low free PSA are more likely to have malignant prostate disease.

Estrogen receptor (ER): ER is a protein found in the nucleus of breast and uterine tissues. The level of ER in the tissue is used to determine whether a person with breast cancer is likely to respond to estrogen therapy with tamoxifen, which binds to the receptors blocking the action of estrogen. Women who are ER-negative have a greater risk of recurrence than women who are ER-positive. Tissue levels are measured using one of two methods. The tissue can be homogenized into a cytosol, and an immunoassay used to measure the concentration of ER receptor protein. Alternatively, the tissue is frozen and thin-sectioned. An immunoperoxidase stain is used to detect and measure the estrogen receptors in the tissue.

Progesterone receptor (PR): PR consists of two proteins, like the estrogen receptor, which are located in the nuclei of both breast and uterine tissues. PR has the same prognostic value as ER, and is measured by similar methods. Tissue that does not express the PR receptors is less likely to bind estrogen analogs used to treat the tumor. Persons who test negative for both ER and PR have less than a 5% chance of responding to endocrine therapy. Those who test positive for both markers have greater than a 60% chance of tumor shrinkage when treated with hormone therapy.

Human chorionic gonadotropin (hCG): hCG is a glycoprotein produced by cells of the trophoblast and developing placenta. Very high levels are produced by trophoblastic tumors and choriocarcinoma. About 60% of testicular cancers secrete hCG. hCG is also produced less frequently by a number of other tumors. Some malignancies cause an increase in alpha and/or beta hCG subunits in the absence of significant increases in intact hCG. For this reason, separate tests have been developed for alpha and beta hCG, and most laboratories use these assays as tumor marker tests. Most EIA tests for pregnancy are specific for hCG, but detect the whole molecule and are called intact hCG assays.

Nuclear matrix protein (NMP22) and bladder tumor-associated analytes (BTA): NMP22 is a structural nuclear protein that is released into the urine when bladder carcinoma cells die. Approximately 70% of bladder carcinomas are positive for NMP22. BTA is comprised of type IV collagen, fibronectin, laminin, and proteoglycan, which are components of the basement membrane that are released into the urine when bladder tumor cells attach to the basement membrane of the bladder wall. These products can be detected in urine using a mixture of antibodies to the four components. BTA is elevated in about 30% of persons with low-grade bladder tumors and over 60% of persons with high-grade tumors.

5.1 Doubling Time Of Tumor Cells

Because cel cycle controles exerted by the Rb, p53, and cyclins are deranged in many tumors, cells can be triggered in to cycle more readily and without the usual restraints. In reality the total cell cycle time for many tumors is equal to or longer than that of corresponding normal cells.

5.2 Growth Fraction

The growth fractionis the proportion of cells within the tumor cells population that are in the replicative pool. During the early, submicroscopic phase of tumor growth, most transformed cells are in the proliferative pool. As tumor continus to grow, cells leave the replicative pool in even- increasing numbers owing to shedding or lack of nutrients, by differentiating, and by reversionto Go. Most cells within cancers remainin the Go phase.

Thus, by the time a tumor is clinically detectable, most cells are not in the replicative pool. Even in some rapidly growing tumors, the growth fraction is approximately 20%.

5.3 Cell Production and Loss

Ultimately, the progressive growth of tumors and the rate at which they grow is determined by how much cell production exceeds cell loss. In sometumors, especially those with a the relatively high growth fraction, the imbalance is large, resulting in more rapid growth than in those in which cell production exceeds cell loss by only a small margin.

An understanding of tumor cell kinetics has important clinical implications.

5.4 Cancer Chemotherapy

Almost all antineoplastic agents in current use are most effective on cycling cells. Hence tumor with high growth fractions are very susceptible to anti-cancer agents in such Cases the tretment strategy is to first shift tumor cells from Go in to the cell cycle. This can be accomplished by debulking the tumor by surgery or radiation. The surviving tumor cells tend to reenter the cell cycle and thus become suceptible to drug therapy.

5.5 Latnt Period of Tumors

If all descendants of an originally transformed cell remained in the replicative pool, most tumors will become clinically detectable within a few months after the first cell division; however most tumor cells leave the replicative pool, therefore the accumuation of cells is a relatively slow process.this in turn rsults in a latent period of sevralmonths to years before a tumor becomes clinically detectable.

The growth of cancer is closely dependent on the balance between cell growth and cell death. Fas (CD95/APO-1), a transmembrane protein related to the TNF-R/NGF-R family, have been shown to be part of a major effector pathway involved in the regulation of apoptosis. Fas are activated by its ligand Fas, which results in the induction of apoptosis. Recently, it was reported that the Fas receptor is not highly expressed in oral squamous carcinomas although the ligand Fas was high in oral carcinomas, particularly in poorly differentiated carcinomas.

Apoptosis in oral squamous carcinoma is lower in poorly differentiated carcinomas, but it is the result of loss of Fas expression or increased anti-apoptotic factors. The Bcl-2 family of proteins appears to regulate apoptosis via differential homodimerisation and heterodimerisation. Bcl-2–Bax heterodimers are an important anti-apoptotic moiety, whereas Bax–Bax homodimers promote cell death. Few studies have been undertaken to investigate apoptosis in oral cancers, although Jordan and colleagues demonstrated that Bcl-2 was present in poorly differentiated cancers, whereas Bax was present in differentiated oral cancers. In a more recent study in oral cancers from an Asian population, low concentrations of Bax were demonstrated, with a high concentration of Bcl-2, irrespective of tumor differentiation. In this regard, it is interesting that recently it is observed in oral primary and metastatic squamous carcinomas an increased expression of $Bclx_s$, Bik, and Bax—proteins that stimulate apoptosis—in contrast to those that inhibit apoptosis, including Bcl_2, Mcl-l, and Bcl-x.

6

Tumor Marker & Carcinogenesis

Neoplasia literally means the process of "new growth" and the new growth is called a neoplasm. Although all physicians know what they mean when they use the term neoplasm, it has been surprisingly difficult to develop an accurate definition. The eminent British oncologist Willis has come closest: "A neoplasm is an abnormal mass of tissue, the growth of which exceeds and is uncoordinated with that of the normal tissues and persists in the same excessive manner after cessation of the stimuli which evoked the change".

Benign tumors are designated by attaching the suffix–*oma* to the cell of origin. Tumors of mesenchymal cells generally follow this rule e.g. a benign tumor arising from fibroblastic cells is called a fibroma, a cartilaginous tumor is a chondroma, and tumor of osteoblast is an osteoma. Nomenclature of benign epithelial tumors is more complex. They based on their cells of origin, on microscopic architecture and others on their macroscopic patterns.

Benign tumors are well differentiated; structure may be typical of tissue origin. Their rate of growth usually progressive and slow in nature, mitotic figures is rare and normal. Nearly all benign tumor grow as cohesive expansile masses that remain localized to their site of origin and do not have the capacity to infiltrate, invade, or metastasize to distant sites.

Malignant tumors arising in mesenchymal tissue are usually called sarcomas because they have little connective tissue stroma and are so fleshy (e.g fibrosarcoma, liposarcoma for smooth muscle cancer and rhabdomyosarcoma for a cancer that arises from striated muscle).

Malignant neoplasms of epithelial cell origin, derived from any of the three germ layers are called carcinomas which can be further classified. One with a glandular growth pattern microscopically is termed as adenocarcinomas and one producing recognizable squamous cells arising in any epithelium of the body is termed as squamous cell carcinoma. In benign and in differentiated malignant neoplasms the parenchymal cells bear a close resemblance to each other were derived from a single cell. Infrequently, a divergent

Manjul Tiwari (MDS), Tumor Marker & Carcinogenesis, 29–40.

differentiation of a single line of parenchymal cells into another tissue are called mixed tumors e.g. mixed tumor of salivary gland origin.

Malignant neoplasms range from well differentiated to undifferentiated, structure is often atypical. Their rate of growth is erratic and may be slow to rapid, mitotic figures may be numerous and abnormal. It is locally invasive, infiltrating the surrounding normal tissues sometimes may be seemingly cohesive and expansile. Metastasis is frequently present, the larger and more undifferentiated the primary, the more likely are metastases.

Oral squamous cell carcinoma arises as a consequence of multiple molecular events induced by the effects of various habits such as tobacco and alcohol use, influenced by environmental factors, possibly viruses in some instances, against a background of inherited resistance or susceptibility. Damage affects many chromosomes and genes, particularly oncogenes, and tumor suppressor genes, and it is the accumulation of such genetic damage and the consequent disturbed cell growth and control which in some instances appear to lead to carcinoma.

Oral squamous carcinogenesis is a multistep process in which multiple genetic events occur that alter the normal functions of oncogenes and tumor suppressor genes. This can result in increased production of growth factors or numbers of cell surface receptors, enhanced intracellular messenger signaling, and/or increased production of transcription factors. In combination with the loss of tumor suppressor activity, this leads to a cell phenotype capable of increased cell proliferation, with loss of cell cohesion, and the ability to infiltrate local tissue and spread to distant sites. Recent advances in the understanding of the molecular control of these various pathways will allow more accurate diagnosis and assessment of prognosis, and might lead the way for more novel approaches to treatment and prevention.

Oral carcinogenesis is a multistep process in which genetic events lead to the disruption of the normal regulatory pathways that control basic cellular functions including cell division, differentiation, and cell death. Several studies have shown that there is a genetic component in the development of carcinoma.

Genetic alterations known to occur during carcinogenesis including point mutations, amplifications, rearrangements, and deletions. Point mutations (single base changes) can lead to over activity or inactivity of gene products. These are common in genes such as K-ras and p53. Amplifications and rearrangements have frequently been reported in malignant neoplasms, with both amplification and rearrangement affecting excitatory pathway genes, whereas rearrangement can also inactivate inhibitory pathway genes.

Several studies have identified specific genetic alterations in oral carcinomas and in premalignant lesions of the oral cavity. Recently, using comparative genomic hybridization on 50 primary head and neck carcinomas, Bockmuhl and colleagues reported deletions of chromosome 3p, 5q, and 9p with 3q gain in well differentiated tumors, whereas in poorly differentiated tumors deletions of 4q, 8p, 11q, 13q, 18q, and 21q and gains in 1p, 11q, 13, 19, and 22q were identified, thus suggesting an association with tumor progression. With the development of molecular techniques, such as micro satellite assays and restriction fragment length polymorphism, it has been shown that allelic imbalance of chromosomal 9p is the most common chromosomal arm loss in head and neck squamous cell carcinoma.

Loss of heterozygosity (LOH) was reported at 9p21–p22 in 72% of tumors. More recently, Partridge and colleagues identified five areas in the region of allelic imbalance at chromosome 3p that might harbor tumor suppressor genes, along with two areas at 8p and 9p, respectively. These authors also identified significantly greater allelic imbalance in patients with TNM stage 4 diseases compared with stages 1–3. Allelic imbalance at one or more loci within 3p24–26, 3p21, 3p13, and 9p21 was associated with reduced survival, with a 25 fold increase in mortality rate with allelic imbalance at 3p24–26, 3p21, and 9p21 compared with patients retaining heterozygosity at these loci.

Allelic loss of 3p and 9p and other regions containing tumor suppressor genes has also been reported in precursor lesions of oral cancer showing varying degrees of dysplasia compared with normal epithelium. Allelic loss or imbalance at p53, DCC (deleted in colon carcinoma), and regions at 3p21.30–22, and 3p12.1–13 were reported, with LOH at DCC shown to occur in areas of dysplasia adjacent to infiltrating carcinoma. This suggested that loss at this locus might be a later event, whereas LOH at 3p and p53 were more frequent in those dysplastic Chromosome breakpoints are frequently seen in centromeric regions of chromosomes 1, 3, 8, 14, 15, 1p22, 11q13, and 19p13. Because genes bcl-1, int-2, and hst-1 have been mapped to 11q13 and n-ras to 11q13, it has been suggested that activation of these oncogenes is the result of these cytogenic alterations.

Approximately two thirds of all head and neck cancers contain a deleted region in chromosome 9p21–22. The cyclin dependent kinases inhibitor 2/multiple tumor suppressor gene 1 (CDKN2/MTSI) has been mapped to this chromosome region, and inactivation of its protein product p16^{INK4} by mutation and deletion has been found in 10% and 33% of head and neck squamous carcinomas, respectively, along with frequent inactivation of p16 in oral premalignant lesions. This suggests an important role for this gene in the early stages of oral carcinogenesis. Cyclins, cyclin dependant kinases (CDKs), and cyclin dependent kinases inhibitors regulate progress through key transitions in the cell cycle. p16^{INK4} binds to and inhibits phosphorylation of pRb by the cyclin dependent kinases CDK4 and CDK6.

Other proteins that regulate crucial checkpoints in the cell cycle, and which are important contributors to increased cell proliferation, include cyclin D, E, and A, which regulate the G1 to S phase transition, and cyclin B, which regulates the G2 to M transition.

The cyclin D1 gene is frequently over expressed in oral cancers as a result of amplification of the 11q13 region. Overexpressin of cyclin a has been reported in oral carcinomas, with the increase in expression being associated with tumor grade. Cyclin B was also reported to be over expressed, with increased cytoplasmic staining compared with nuclear staining in normal cells. Cyclin B1 binds to protein kinases p34^{cdc2} in the cytoplasm of the dividing cells, and the complex is transported to the nucleus at the G2 to M transition. This suggests that frequent abnormalities in cyclin B/p34^{cdc2} kinetics in oral carcinomas lead to deregulation of the G2 to M transition.

During oral carcinogenisis, intracellular messengers might also be intrinsically activated, thereby overriding the necessity for ligand–receptor regulated signals. Among the genes involved in intracellular signaling pathways, members of the ras family have been examined in oral cancers. H-ras, K-ras, and N-ras all encode the protein p21, which is located on the cytoplasmic membrane of the cell membrane, and which transmits mitogenic signals by binding GTP. The mitogenic signal is terminated by the conversion of GTP to GDP by hydrolysis, but when the ras oncogene is mutated this conversion can be prevented, thus leading to continuous stimulation.

Some studies have indicated that members of the ras oncogene family are over expressed in oral cancers. Although loss of control of N-ras might be an early step in carcinogenisis in oral cancers, with increased expression occurring early in dysplastic lesions, ras mutations are uncommon in the progression of oral cancers in the Western world, occurring in less than 5% of all cases. In contrast, 55% of lip cancers have H-ras mutation and H-ras mutation occurs in 35% of oral cancers in the Asian population, where it is especially associated with betel nut chewing.

Transcription factors that activate other genes are also activated in oral cancer. The functional activity of many of these proteins is regulated by receptor activated second messenger pathways, and neutralization of these genes could result in a cell cycle block, preventing mitogenic and differentiation responses to growth factors. Among these genes c-myc, which helps regulate cell proliferation, is frequently over expressed in oral cancers as a result of gene amplification. Over expression is frequently associated with poorly differentiated tumors, although more recently c-myc has been shown to be over expressed in moderate and well differentiated oral carcinomas, in which cell proliferation far outweighed the number of apoptotic cells present (HK Williams *et al.*, unpublished data, 1999). C-Myc induces both cell proliferation and apoptosis. C-Myc requires p53 to induce apoptosis and the retinoblastoma tumor suppressor gene Rb-1 nuclear protein pR6 interacts with the c-myc gene, preventing its transcription, and thus inhibiting cell proliferation. However, on phosphorylation of pR6, c-Myc is increased and cell proliferation proceeds. In this regard, it is interesting that we found pR6, c-myc, and p53 to be expressed in all oral carcinomas, irrespective of differentiation. However, further studies are needed to determine the genetic status of these oncogenes and tumor suppressor genes, and to measure the concentrations of the proteins, to determine possible controlling pathways in the development of these oral cancers (HK Williams *et al.*, unpublished results, 1999).

The PRAD-1 gene located on 11q13 encodes cyclin D, which together with the Rb gene product controls the G1 to S transition of the cell cycle. The PRAD-1 gene is amplified in 30–50% of head and neck cancers. Amplification of PRAD-1 is correlated with cytological grade, infiltrative growth pattern, and metastases.

The hst-1/int-2 gene encodes a protein that is homologous to fibroblast growth factor, and which in oral cancers has been shown to be involved in tumor growth, and to have angiogenic activity. This gene maps to human chromosome 11q13.3, which is co amplified with int-2 in some cancers. Lese *et al.*, in 1995, reported co amplification of the int-2 and hst-1 genes in oral squamous carcinomas, and other authors have suggested that this co amplification in head and neck squamous carcinomas is associated with tumor recurrence and progression of disease.

Int-2 amplification has also been described in premalignant lesions adjacent to neoplasia, both in areas of dysplasia and hyperplasia, which suggests that int-2, can be amplified before tumor development.

6.1 Malignant Tumors

Tissue of Origin   Malignant
Composed of one Parenchymal
Cell type

Tumors of mesenchymal origin

Connective tissue and derivatives	Fibrosarcoma
	Liposarcoma
	Chondrosarcoma
	Osteogenic sarcoma

Endothelial and related tissues

Blood vessels	Angiosarcoma
Lymph vessels	Lymphyangiosarcoma
Synovium	Synovial sarcoma
Mesothelium	Mesothelioma

Blood cells and related cells

Hematopoietic cells	Leukemias
Lymphoid tissue	Lymphomas

Muscle

Smooth	Leiomyosarcoma
Striated	Rhabdomyosarcoma

Tumors of epithelial origin

Stratified squamous	Squamous cell or epidermoid Carcinoma
Basal cells of skin	Basal cell carcinoma
Epithelial lining of glands or ducts	Adenocarcinoma
	Papillary carcinomas
	Cystadenocarcinoma
Salivary glands	Malignant mixed tumor of Salivary gland origin[48]

6.1.1 Squamous cell originate from malignant tumors

1. Verrucous carcinoma
2. Spindle cell carcinoma
3. Adenosquamous carcinoma
4. Basaloid squamous carcinoma
5. Carcinoma of maxillary sinus
6. Nasopharyngeal carcinoma
7. Keratoacantoma[6]

Cancer of the oral cavity is more prevalent in developing countries, where many people are addicted to tobacco chewing and maintain poor oral hygiene.

Among the more pressing problems in clinical management is the lack of early detection, due to the absence of a potential diagnostic marker. Oncologists are now more aware of the challenges associated with the treatment of cancer of the oral cavity, and survival percentages are improving significantly.

> According to the WHO agreement of 1973, oral cancer is a malignant neoplasm in the 8 anatomic regions of the oral cavity. Thus oral cancer refers to cancerous tumors of the upper respiratory and alimentary tracts and in the near vicinity, whose draining lymphatic vessels are all located in the neck. Since the treatment of malignant disease which has spread to the lymphatic system entails treating the lymphatics and the primary tumors one entity, the therapeutic plan must include the neck region. Thus the definition of oral cancer also applies to so-called neck and head tumors,'

Quote from the inaugural speech of Prof. Wilfred Schilli from the University of Freiburg, Germany, in 1989.

Carcinogenesis is a multistep process at both the phenotypic and genetic level. A malignant neoplasm has several phenotypic attributes, such as excessive growth, local invasiveness, and the ability to form distant metastasis. These characteristics are acquired in a stepwise fashion, a phenomenon called tumor progression.

At the molecular level, progression results from accumulation of genetic lesions that in some instances are favored by defects in DNA repair.

More than one mutation is necessary for carcinogenesis. Only mutations in those certain types of genes which play vital role in cell division, cell death, and DNA repair will cause a cell to lose control of its proliferation.

Carcinogenesis is caused by mutation of the genetic material of normal cells, which upsets the normal balance between proliferation and cell death. This results in uncontrolled cell division and tumor formation.

Oral carcinogenesis is a multistep process in which genetic events lead to the disruption of the normal regulatory pathways that control basic cellular functions including cell division, differentiation, and cell death.

There is a genetic component in the development of oral carcinoma. These include the occurrence of familial aggregations of cancer, including oral cancer, with carcinomas developing at a younger age. Whether patients develop single site oral cancer or multiple site oral cancer, much evidence has accumulated to suggest that multiple genetic events lead to oral cancer, with around six to 10 genetic events believed to result in oral carcinogenesis. Genetic alterations known to occur during carcinogenesis including point mutations, amplifications, rearrangements, and deletions. Point mutations (single base changes) can lead to over activity or inactivity of gene products. These are common in genes such as K-ras and p53. Amplifications and rearrangements have frequently been reported in malignant neoplasms, with both amplification and rearrangement affecting excitatory pathway genes, whereas rearrangement can also inactivate inhibitory pathway genes.

Oral squamous cell carcinoma arises as a consequence of multiple molecular events induced by the effects of various habits such as tobacco and alcohol use, influenced by environmental factors, possibly viruses in some instances, against a background of inherited resistance or susceptibility. Damage affects many chromosomes and genes, particularly oncogenes, and tumor suppressor genes, and it is the accumulation of such genetic damage and the consequent disturbed cell growth and control which in some instances appear to lead to carcinoma.

Oral squamous carcinogenesis is a multistep process in which multiple genetic events occur that alter the normal functions of oncogenes and tumor suppressor genes. This can result in increased production of growth factors or numbers of cell surface receptors, enhanced intracellular messenger signaling, and/or increased production of transcription factors. In combination with the loss of tumor suppressor activity, this leads to a cell phenotype capable of increased cell proliferation, with loss of cell cohesion, and the ability to infiltrate local tissue and spread to distant sites. Recent advances in the understanding of the molecular control of these various pathways will allow more accurate diagnosis and assessment of prognosis, and might lead the way for more novel approaches to treatment and prevention.

Squamous cell carcinoma of the head and neck is the sixth most common human malignancy, although it only accounts for 2% of all cancers in Western populatios. However, the incidence of head and neck cancer, in particular tumors of the larynx and oral cavity, are increasing in developed countries, with the increase of risk being seen in younger people, particularly young women. Despite therapeutic changes survival remains poor, mainly because of the increased risk of developing a second malignancy, which can be a second primary carcinoma.

Oral carcinogenesis is a multistep process in which genetic events lead to the disruption of the normal regulatory pathways that control basic cellular functions including cell division, differentiation, and cell death. Several studies have shown that there is a genetic component in the development of carcinoma.

Genetic alterations known to occur during carcinogenesis including point mutations, amplifications, rearrangements, and deletions. Point mutations (single base changes) can lead to over activity or inactivity of gene products. These are common in genes such as K-ras and p53. Amplifications and rearrangements have frequently been reported in malignant neoplasms, with both amplification and rearrangement affecting excitatory pathway genes, whereas rearrangement can also inactivate inhibitory pathway genes.

Several studies have identified specific genetic alterations in oral carcinomas and in premalignant lesions of the oral cavity. Recently, using comparative genomic hybridization on 50 primary head and neck carcinomas, Bockmuhl and colleagues reported deletions of chromosome 3p, 5q, and 9p with 3q gain in well differentiated tumors, whereas in poorly differentiated tumors deletions of 4q, 8p, 11q, 13q, 18q, and 21q and gains in 1p, 11q, 13, 19, and 22q were identified, thus suggesting an association with tumor progression. With the development of molecular techniques, such as micro satellite assays and restriction fragment length polymorphism, it has been shown that allelic imbalance of chromosomal 9p is the most common chromosomal arm loss in head and neck squamous cell carcinoma.

Loss of heterozygosity (LOH) was reported at 9p21–p22 in 72% of tumors. More recently, Partridge and colleagues identified five areas in the region of allelic imbalance at chromosome 3p that might harbor tumor suppressor genes, along with two areas at 8p and 9p, respectively. These authors also identified significantly greater allelic imbalance in patients with TNM stage 4 diseases compared with stages 1–3. Allelic imbalance at one or more loci within 3p24–26, 3p21, 3p13, and 9p21 was associated with reduced survival, with a 25 fold increase in mortality rate with allelic imbalance at 3p24–26, 3p21, and 9p21 compared with patients retaining heterozygosity at these loci.

Allelic loss of 3p and 9p and other regions containing tumor suppressor genes has also been reported in precursor lesions of oral cancer showing varying degrees of dysplasia compared with normal epithelium. Allelic loss or imbalance at p53, DCC (deleted in colon carcinoma), and regions at 3p21.30–22, and 3p12.1–13 were reported, with LOH at DCC shown to occur in areas of dysplasia adjacent to infiltrating carcinoma. This suggested that loss at this locus might be a later event, whereas LOH at 3p and p53 were more frequent in those dysplastic Chromosome breakpoints are frequently seen in centromeric regions of chromosomes 1, 3, 8, 14, 15, 1p22, 11q13, and 19p13.

Because genes bcl-1, int-2, and hst-1 have been mapped to 11q13 and n-ras to 11q13, it has been suggested that activation of these oncogenes is the result of these cytogenic alterations.

Approximately two thirds of all head and neck cancers contain a deleted region in chromosome 9p21–22. The cyclin dependent kinases inhibitor 2/multiple tumor suppressor gene 1 (CDKN2/MTSI) has been mapped to this chromosome region, and inactivation of its protein product p16^{INK4} by mutation and deletion has been found in 10% and 33% of head and neck squamous carcinomas, respectively, along with frequent inactivation of p16 in oral premalignant lesions. This suggests an important role for this gene in the early stages of oral carcinogenesis. Cyclins, cyclin dependant kinases (CDKs), and cyclin dependent kinases inhibitors regulate progress through key transitions in the cell cycle. p16^{INK4} binds to and inhibits phosphorylation of pRb by the cyclin dependent kinases CDK4 and CDK6.

Other proteins that regulate crucial checkpoints in the cell cycle, and which are important contributors to increased cell proliferation, include cyclin D, E, and A, which regulate the G1 to S phase transition, and cyclin B, which regulates the G2 to M transition.

The cyclin D1 gene is frequently over expressed in oral cancers as a result of amplification of the 11q13 region. Overexpressin of cyclin a has been reported in oral carcinomas, with the increase in expression being associated with tumor grade. Cyclin B was also reported to be over expressed, with increased cytoplasmic staining compared with nuclear staining in normal cells. Cyclin B1 binds to protein kinases p34^{cdc2} in the cytoplasm of the dividing cells, and the complex is transported to the nucleus at the G2 to M transition. This suggests that frequent abnormalities in cyclin B/p34^{cdc2} kinetics in oral carcinomas lead to deregulation of the G2 to M transition.

Oral carcinoma is the most common malignancy of orofacial region though strictly speaking carcinoma means malignancies of epithelial tissue origin.

The malignant neoplasms of epithelial cell origin, derived from any of the three germ layers, are called carcinomas. Thus cancer arising in the epidermis of ectodermal origin is a carcinoma, as is a cancer arising in the mesodermally derived cells of the renal tubules and the endodermally derived cells of the lining of gestrointestinal tract.

Carcinomas may be further qualified. One with a glandular growth pattern microscopically is termed an adenocarcinoma, and one producing recognizable squamous cells arising in any epithelium of the body is termed a squamous cell carcinoma.

Carcinogenesis is a multistep process at both phenotypic and genetic level means the creation of cancer. It includes the process of derangement of the rate of cell division due to damage to D.N.A. so cancer is ultimately a disease of genes. Cancer is caused by a series of mutations. Each mutation alters the behavior of the cell.

The transformation of normal cells in to malignant cells is dependent on mutations in the genes that control cell cycle progression, leading to the loss of regulatory cell cycle growth signals.

It might then be profitable to list some fundamental principles before we delve in to the details of the molecular basis of cancer.

At the molecular level, progression results from accumulation of genetic lesions that in some instances are favored by defects in D.N.A repair.

Three classes of normal regulatory genes-the growth promoting proto-oncogene, the growth inhibiting cancer suppressor genes(antioncogenes), and genes that regulate programmed cell death, or apoptosis-are the principal targets of genetic damage.

Among the molecular mechanisms involved in the carcinogenesis, defects in the regulation of programmed cell death (apoptosis) may contribute to the pathogenesis and progression of cancer. Dysregulation of oncogenes and tumor suppressor genes involved in apoptosis are also associated with tumor development and progression.

Genes that regulate apoptosis may be dominant, as are proto-oncogene, or they may behave as cancer suppressor genes.

In addition to the three classes of genes mentioned earlier, a fourth category of genes, those that regulate repair of damaged DNA are also pertinent in carcinogenesis.

DNA repair genes affect cell proliferation or survival indirectly by influencing the ability of the organism to repair nonlethal damage in other genes, including proto-oncogenes, tumor suppressor genes, and genes that regulate apoptosis.

The unregulated growth that characterizes cancer is caused by damage to DNA, resulting in mutations to gene that encode for proteins controlling cell division. Many mutation events may be required to transform a normal cell in to a malignant cell. These mutations can be caused by chemicals or physical agents called carcinogens, by close exposure to radioactive materials or by certain viruses that can insert their DNA in to the human genome.

Mutations occur spontaneously, and may be passed down from one generation to the next as a result of mutations within germ line.

The genetic hypothesis of cancer implies that a tumor mass results from the clonal expansion of a single progenitor cell that has incurred the genetic damage. Clonality of tumors is assessed quite readily in women who are heterozygous for polymorphic X-linked markers, such as the enzyme glucose-6-phosphate dehydrogenase (G6PD) or X-linked restriction fragment length polymorphism.

A malignant neoplasm has several phenotypic attributes, such as excessive growth, local invasiveness, and the ability to form distant metastases. These characteristics are acquired in a stepwise fashion, a phenomenon called tumor progression.

Many forms of cancer are associated with exposure to environmental factors such as tobacco smoke, radiation, alcohol and certain viruses.

A variety of agents increase the frequency with which cells are converted to the transformed condition, they are said to be carcinogenic agents. Carcinogens may cause epigenetic changes or may act directly or indirectly to change the genotype of the cells.

Although tobacco is clearly of major aetiological significance (IARC 1984) the failure of overtly malignant lesions to develop in all tobacco users and the development of oral cancer in all tobacco users and the development of oral cancer in persons with no history of tobacco use suggests that the genesis of oral cancer may also involve other unidentified environmental and host factors.

Several studies have identified specific genetic alterations in oral carcinomas and in premalignant lesions of the oral cavity. Recently, using comparative genomic hybridization on primary oral carcinomas, Bockmuhl and colleagues reported deletions of chromosome 3p, 5q, and 9p with 3q gain in well differentiated tumors, whereas in poorly differentiated tumors deletions of 4q, 8p, 11q, 13q, 18q, and 21q and gains in 1p, 11q, 13, 19, and 22q were identified, thus suggesting an association with tumor progression. With the development of molecular techniques, such as micro satellite assays and restriction fragment length polymorphism, it has been shown that allelic imbalance of chromosomal 9p is the most common chromosomal arm loss in oral carcinoma.

Several oncogenes have also been implicated in oral carcinogenesis. Aberrant expression of the proto-oncogene epidermal growth factor receptor (EGFR/c-erb 1), members of the ras gene family, c-myc, int-2, hst-1, PRAD-1, and bcl-1 is believed to contribute towards cancer development.

7

Management

Every year the approach to laboratory diagnosis of cancer becomes more complex, more sophisticated, and more specialized.

The laboratory evaluation of a lesion can be only as good as the specimen made available for examination. It must be adequate, representative and properly preserved. Several sampling approaches are available:

1. BIOPSY: It is a deliberate & controlled process of collection of tissue from living individual for examination, histopathological evalution, microbial analysis, & biochemical assessment or combination of all the above.
2. Fine –needle aspiration of tumors is another approach that is widely used. The procedure involves aspirating cells and attendant fluid with a small –bore needle, followed by cytologic examination of the stained smear.

The tissue/cytological smear hence collected can further be processed and stained or exposed to advanced molecular diagnostic methods to achieve final diagnosis.

1. Routine H&E staining: Most widely used and important general purpose stain combination. May be used after any fixation except fixation with osmium tetraoxide. Hematoxylin, a natural dye product, acts as a basic dye that stains blue or black. Nuclear heterochromatin stains blue and the cytoplasm of cells rich in ribonucleoprotein also stains blue. The cytoplasm of cells with minimal amounts of ribonucleoprotein tends to be lavender in color.

   The aniline dye, eosin, is an acid dye that stains cytoplasm, muscle, and connective tissues various shades of pink and orange. This difference in staining intensity is useful in differentiating one tissue from another. Although it is an esthetically pleasing combination and widely used, it is limited in its ability to differentiate cytoplasmic organelles.

Manjul Tiwari (MDS), Tumor Marker & Carcinogenesis, 41–47.

2. Immunohistochemistry

The availability of specific monoclonal antibodies has greatly facilitated the identification of cell products or surface markers. It can be employed for:

(a) Catergorization of undifferentiated malignant tumors: In many cases malignant tumors of diverse origin resemble each other because of poor differentiation. These tumors are often quite difficult to distinguish on the basis of routine hematoxylin and eosin – stained tissue sections. Antibodies against intermediate filaments have proved to be value in such cases because tumor cells often contain intermediate filaments characterized of their cell of origin.

(b) Determination of site of origin of metastatic tumors: Many cancer patients present with metastases. In cases in which the origin of the tumor is obscure, immunohistochemical detection of tissue specific or organ specific antigens in a biopsy specimen of the metastatic deposit can lead to the identification of the tumor source.

The ret proto-oncogene, a receptor tyrosine kinase, exemplifies oncogenic conversion via mutations and gene rearrangements.

Point mutations in the ret proto-oncogene are associated with dominantly inherited MEN types 2A and 2B. In MEN 2A, point mutations in the extra cellular domain cause constitutive dimerization and activation, whereas in MEN 2B, point mutations in the cytoplasmic catalytic domain activate the receptor. In all these familial tumors, the affected individuals inherit the ret mutations in the germ line.

In myeloid leukemias, the gene encoding the colony stimulating factor1 (CSF1) receptor has been detected. In certain chronic myelomonocytic leukemias with the t (12:9) translocation, the entire cytoplasmic domain of the PDGF receptor is fused with a segment of the ETS family transcription factor, resulting in permanent dimerization of the PDGF receptor.

Three members of the EGF receptor family are the ones most commonly involved. The normal form of c-erb B1, the EGF receptor gene, is over expressed in squamous cell carcinomas of lung, and less commonly in carcinomas of urinary bladder, gastrointestinal tract and astrocytomas. This increased receptor expression results from gene amplification. The c-erb B2 gene (also called c-neu), the second member of the EGF receptor family, is amplified in a high percentage of human Aden carcinomas arising within the breast, ovary, lung, stomach and salivary glands. A third member of the EGF receptor family, c-erb B3 is also over expressed in breast cancers.

3. Molecular Diagnosis

Several molecular techniques have been used for diagnosis and in some cases for predicting behavior of tumors.

Diagnosis of hereditary predisposition to cancer: Germ line mutations in several tumor suppressor genes which are associated with a high risk of developing specific cancers. Thus detection of carriers of these mutations in family members of affected patients or in those at high risk of carrying the mutation has become important. Such analysis usually requires detection of a specific mutation or sequencing of the entire gene.

In most cases, oral cancer can only be diagnosed by a biopsy (removal of tumor cells for viewing under a microscope). But markers can help determine if a cancer is likely. They can also help diagnose the source of widespread cancer in a patient when the origin of the cancer is unknown.

Changes in some tumor markers have been sensitive enough to be used as targets in clinical trials. When used for diagnosis, tumor markers are used in conjunction with other clinical parameters.

1. Prognosis of malignant neoplasms: Certain genetic alterations are associated with poor prognosis and hence their detection allows stratification of patients for therapy. These can be detected by routine cytogenetics and also by FISH or PCR assays.
2. Detection of minimal residual disease: After treatment of patients with leukemia or lymphoma, the presence of minimal disease or the onset of relapse can be monitored by PCR –based amplification of unique nucleic acid sequences generated by translocation.
3. DNA microarray analysis and proteomics: These methods are used to obtain gene expression signatures (molecular profiles) of cancer cells. DNA microarray techniques reveal the RNA expression from as many as 30,000 different genes using gene chip technology.

Some newer tumor markers help show how aggressive a particular oral cancer is, or even how well it might respond to a particular drug.

Detection of molecules that have therapeutic significance: Immunohistochemical detection of hormone (estrogen/progesterone) receptors in breast cancer cells is of therapeutic value because these cancers are susceptible to antiestrogen therapy.

In any oral cancer if the marker level in the blood goes down, it is almost always a sign that the treatment is having an effect. On the other hand, if the marker level goes up, then the treatment probably should be changed. Although there are a multitude of tumor markers, very few of them have found their way into clinical practice because of their lack of specificity. However, some of these non-specific markers have found a place in monitoring cancer treatment rather than in diagnosis.

The signal transducing proteins are oncoproteins that mimic the function of normal cytoplasmic signal transducing proteins.

Such proteins are strategically located on the inner leaflet of the plasma membrane, where they receive signals from outside the cell by activating GFR and transmit them to the cells nucleus. The best and most well studied example of a signal transducing oncoprotein is ras family of guanine triphosphate (GTP) binding proteins.

Most such proteins are strategically located on the inner leaflet of the plasma membrane, where they receive signals from outside the cells (e.g. by activation of growth factor receptors) and transmit them to the cell nucleus. The signal transducing proteins are heterogeneous. The best example of a signal transducing oncoprotein is the ras family of guanine triphosphate (GTP) - binding protein.

The ras proteins were discovered initially in the form of viral oncogenes. Approximately 10 to 20% of all human tumors contain mutated versions of ras proteins.

Mutation of the gene is the single most common abnormality of dominant oncogenes in human tumors. ras plays an important role in mutagenesis induced by growth factors for example, blockade of ras function by microinjection of specific antibodies blocks the proliferative response to EGF, PDGF, and CSF-1.

In the inactive state, ras proteins bind guanosine diphosphate (GDP). When cells are stimulated by growth factors or other receptor ligand interactions, ras becomes activated by exchanging GDP for GTP. The activated ras excites the MAP kinase pathway by recruiting the cytosolic protein raf-1. The MAP kinases activated target nuclear transcription factors and thus promote mitogenesis.

In normal cells, the activated signal transmitting stage of ras protein is transient because its intrinsic GTPase activity hydrolyzes GTP to GDP, thereby returning ras to its quiescent ground state.

The orderly cycling of the ras protein depends on two reactions:

1. Nucleotide exchange (GDP by GTP) which activates ras protein, and
2. GTP hydrolysis, which converts the GTP-bound inactive form.

The removal of GDP and its replacement by GTP during ras activation is catalyzed by a family of guanine nucleotide releasing proteins that are recruited to the cytosolic aspect of activated growth factor receptors by adaptor proteins. The GTPase activity is accelerated by GTPase activating proteins (GAPs). These widely distributed proteins bind to the active ras and augment its GTPase activity by more than 1000-fold leading to rapid hydrolysis of GTP to GDP and termination of signal transduction. Thus, GAPs function as 'brakes' that prevent uncontrolled ras activity.

Many of these proteins bind to DNA at specific sites from which they can activate or inhibit transcription of adjacent genes.

A whole host of oncoproteins, including products of the myc, myb, Jun, and fos oncogenes, have been localized to the nucleus of these myc gene is most commonly involved in human tumors.

The c-myc proto-oncogene is expressed in virtually all eukaryotic cells and belongs to the immediate early growth response genes, which are rapidly induced when quiescent cells receive a single to divide.

After translation, c-myc protein is rapidly translocated to the nucleus. Either before of after transport to the nucleus, it forms a heterodiamer with another protein called max. The myc-max heterodimer binds to specific DNA sequences (termed E-boxes) and is a potent transcriptional activator. Mutations that impair the ability of myc to bind to DNA or to max also abolish its oncogenic activity.

Mad another member of the myc super family of transcriptional regulators, can also bind max to form a dimer. The max-mad heterodimer functions as a transcription repressor. Thus emerging theme seems to be that the degree of transcriptional activation by c-myc is regulated not only by the levels of myc protein but also by the abundance and availability of max and mad proteins. In this network myc-max favors proliferation, whereas mad-max inhibits cell growth. mad may therefore be considered an antioncogene (tumor suppressor gene).

It is becoming increasingly evident that myc not only controls the cell growth, but also it can drive cell death by apoptosis. Thus, when myc activation occurs in the absence of survival signals (growth factors) cells undergo apoptosis.

Dysregulation of c-myc expression resulting from translocation of gene occurs in Burkett's lymphoma, c-myc is amplified in breast, colon, lung and many other carcinomas. N-myc and L-myc genes are amplified in neuroblastomas and small cell cancers of lungs. The related N-myc and L-myc genes are amplified in neuroblastomas and small cell cancers.

Various phases of cell cycle are orchestrated by cyclins and cyclin dependent kinases (CDKs) and their inhibitors. Mutations in genes that encode this cell cycle have been found in several human cancers.

Cyclin dependent kinases are expressed constitutively during the cell cycle but in an inactive form. They are activated by phosphorylation after binding to another family of proteins, called cyclins. Cyclins are synthesized during specific phases of the cell cycle, and their function is to activate the CDKs. on completion of this task, cyclin levels decline rapidly. Each phase of the cell cycle circuitry is carefully monitored, the transition from G1 to S is an extremely important checkpoint in the cell cycle clock because once cells cross this barrier they are committed progress in to S phase. When a cell receives growth promoting signals, the synthesis of D type cyclins that bind CDK4 and CDK6 is stimulated in the early part of G1. Later in the G1 phase of the cell cycle, the synthesis of E cyclin is stimulated, which in turn, binds to CDK2. The cyclin D/CDK4, CDK6 and cyclin E/CDK2 complexes phosphorylated with the retinoblastoma proteins (pRb). pRb binds to the E2F family of transcription factors. Phosphorylation of pRb unshackles the E2F proteins and they in turn activate the transcription of several genes whose products are essential for progression through the S phase. These include DNA polymerases, thymidine kinase, and dihydrofolate reductase. The progress of cells from the S phase in to the G2 phase is facilitated by up-regulation of cyclin A which binds to CDK2 and to CDK1.

Early in the G2 phase, B cyclin takes over. By forming complexes with CDK1, it helps the cell move from G2 to M.

The activity of CDKs is regulated by two families of CDK inhibitors (CDKIs). One family of CDKIs composed of three proteins, called p21, p27 and p57, inhibits the CDKs broadly. Whereas the other family of CDKIs has selective effects on cyclin D/CDK4 and cyclin D/CDK6.

The four members of this family (p15, p16, p18, p19) are sometimes called INK4 proteins (because they are inhibitors of CDK4 and CDK6).

Amplification of CDK4 gene occurs in melanomas, sarcomas, and glioblastomas. Mutations affecting cyclin B and cyclin E and other CDKs also occur in certain malignant neoplasms but they are much less frequent than those affecting cyclin D/CDK4.

There has been much research on the tumor suppressor gene p53. The p53 protein blocks cell division at the G1 to S boundary, stimulates DNA repair after DNA damage, and also induces apoptosis. These functions are achieved by the ability of p53 to modulate the expression of several genes. The p53 protein transcriptionally activates the production of the p21 protein, encoded by the WAF1/CIP gene, p21 being an inhibitor of cyclin and cyclin dependant kinase complexes. p21 transcription is activated by wild-type p53 but not mutant p53. However, WAF1/CIP expression is also induced by p53 independent pathways such as growth factors, including platelet derived growth factor(PDGF), fibroblast growth factor(FGF), and transforming growth factor ß(TGF-ß). Wild-type p53 has a very short half life (four to five minutes), whereas mutant forms of protein are more stable, with a six hour half life.

Mutation of p53 occurs either as a point mutation, which results in a structurally altered protein that sequesters the wild-type protein, thereby inactivating its suppressor activity, or by deletion, which leads to a reduction or loss of p53 expression and protein function.

The tumor suppressor gene p53 is known to be mutated in approximately 70% of adult solid tumors.

p53 has been shown to be functionally inactivated in oral tumors, and restoration of p53 in oral cancer lines and tumors induced in animal models has been shown to reverse the malignant phenotype. Smoking and tobacco use have been associated with the mutation of p53 in head and neck cancers.

Other tumor suppressor genes include doc-1, the retinoblastoma gene, and APC. The doc-1 gene is mutated in malignant oral keratinocytes, leading to a reduction of expression and protein function. The precise function of the Doc-1 protein in oral carcinogenisis is unclear, but it is very similar to a gene product induced in mouse fibroblasts by tumor necrosis factor α(TNF-α). Normally, TNF-α decreases proliferation and increases differentiation, and has been shown in oral squamous cell carcinoma cell lines to be responsible, either alone or in combination with interferons α or γ, for antiproliferative activity.

8

Tumor Marker in Relation to Carcinogenesis

The Tumor Markers having oral implications are enlisted as following:

ONCOGENES:
p53

p16^{INK4a} and p14^{ARF}
Cyclin-dependent kinase inhibitors
Retinoblastoma protein (pRb)
EGF (Epithelial Growth Factor) (EGF-R, c-erb1-4 o Her-2/neu)
Cyclins (cyclin A, B1, D1, E)
Bcl2/BAG1
Fas/FasL
Ki-67/MIB

8.1 Tumour Growth Markers

Nuclear cell proliferation antigens
AgNOR (argyrophilic nucleolar organizer region) associated proteins
Skp2 (S-phase kinase-interacting protein 2)
HSP27 and 70 (heat shock proteins)
Telomerase

8.2 Markers of Tumour Suppression and Anti-Tumour Response

Bax
Dendritic cells (DC)
Zeta chains

Manjul Tiwari (MDS), Tumor Marker & Carcinogenesis, 49–125.
© *2012 River Publishers. All rights reserved.*

8.3 Angiogenesis Markers

 (i) *VEGF/VEGF-R (vascular endothelial growth factor/receptor)*
 (ii) *NOS2 (nitric oxide synthase type II)*
(iii) *PD-ECGF (platelet-derived endothelial cell growth factor)*
(iv) *FGFs (fibroblast growth factor)*

8.4 Markers of Tumour Invasion and Metastatic Potential

 (i) *MMPs (matrix-metallo proteases)*
 (ii) *Integrins*
(iii) *Cadherins and catenins*
(iv) *Desmoplakin/placoglobin*
 (v) *Ets-1*

8.5 Cell Surface Markers

 (i) *Carbohydrates and antigen*

8.6 Intracellular Markers

 (i) *Cytokeratins*

8.7 Markers of Anomalous Keratinisation

 (i) *Filagrins*
 (ii) *Involucrin*
(iii) *Desmosomal proteins*
(iv) *Intercellular substance antigen*
 (v) *Nuclear analysis*

8.8 Arachidonic Acid Products

8.8.1 Enzymes

Molecular Markers of the Risk of Oral Cancer

8.8.2 Survivin

8.9 Oncogenes

These cellular genes were first discovered by Noble Laureate Michael Bishop and Harold Varmus as passengers within genome of acute transforming retroviruses, which cause rapid induction of tumors in animals and can also transform animal cells in vitro.

Genes that promote autonomous cell growth in cancer cells are known as oncogenes and their normal cellular counterparts are called protooncogenes. Protooncogenes are physiologic regulators of cell proliferation and differentiation; oncogenes are characterized by the ability to promote cell growth in the absence of normal mitogenic signals. Their products called *oncoprotein*, resemble the normal products of protooncogenes. Protoncogenes, promote cell growth through variety of ways. Many can produce hormones a "Chemical Messenger" between cells which encourage mitosis, the effect of which depends on the signal transduction of the receiving tissue or cells, Some are responsible for the signal transduction system and signal receptors in cells and tissues themselves, thus controlling the sensitivity to such hormones. They often produce mutagens or are involved in transcription of DNA in protein synthesis which creates the proteins and enzymes responsible for producing the products and biochemicals cells use and interact with normal genes or cellular protooncogenes can be activated by structural or functional alteration for example point mutation, gene amplification, proviral insertions, chromosomal rearrangement.

Oncogenes are altered growth promoting regulatory genes that govern the cells signal transduction pathways, and mutation of these genes leads to either overproduction or increased function of the excitatory proteins. Although oncogenes alone are not sufficient to transform epithelial cells, they appear to be important initiators of the process, and are known to cause cellular changes through mutation of only one gene copy.

Several oncogenes have been implicated in oral carcinogenesis. Some important oncogenes are BRAC-1(located on chromosome 17q12-21 is recently discovered tumor suppressor genes that are associated with the occurrence of breast and several other cancers), myc, p53, RB (retinoblastoma) gene (RB), and Ph[1] (Philadelphia chromosome).

The regulation of normal cell growth and division is subject to the action of positive activators, which, in response to the proper stimuli, activate cytoplasmic products or the transcription of specific genes. The normal growth of cells is a balance between regulation by positive activators, which stimulate growth and regulation by negative–acting factors which inhibit growth. In either case, a mutation occurs in a genes coding for these factors can lead to the loss of growth regulation and development of cancer. When mutation occurs in a gene that acts in a negative fashion and the product of the gene is either lost or altered, negative regulation is also lost, and uncontrollable cell growth occurs. Genes whose loss or inactivation leads to uncontrollable cell proliferation are known as tumor suppressor genes. They are important for regulating cell cycle progression, functioning to stop or start the growth of a cell at a particular phase in the cell cycle. They can act by regulating the transcription of specific genes, either by activating transcription of some genes or by repressing transcription of other genes. Tumor suppressor proteins also can direct a cell to undergo differentiation or to undergo a programmed death.

This is brought about by two broad categories of changes-

- Changes in the structure of the gene, resulting in the synthesis of an abnormal gene product (oncoprotein) having an aberrant function.
- Changes in the regulation of gene expression, resulting in enhanced or inappropriate production of the structurally normal growth promoting protein.

The ras oncogenes represent the best example of activation by point mutations. A large number of human tumors carry ras mutations. The frequency of such mutations varies with different tumors, but in some types it is high. A single nucleotide base may be substituted by a different base, resulting in a point mutation. Point mutations reduce the GTPase activity of the ras proteins.

Point mutations (single base changes) can lead to over activity or inactivity of gene products. These are common in genes such as K-ras and p53. Amplifications and rearrangements have frequently been reported in malignant neoplasms, with both amplification and rearrangement affecting excitatory pathway genes, whereas rearrangement can also inactivate inhibitory pathway genes.

In general, carcinomas have mutations of K-ras, whereas hematopoietic tumors bear N-ras mutations. ras mutations are infrequent or even nonexistent in certain other tumors. Therefore although ras mutations are extremely common, their presence is not essential for carcinogenesis. In addition to ras, activating point mutations have been found in the c-fms gene in some cases of acute myeloid leukemia.

Two types of chromosomal rearrangements can activate proto-oncogenes-translocations and inversions. Of these chromosomal translocations are much more common. Translocations can activate proto-oncogenes in two ways.

In lymphoid tumors, specific translocations result in over expression of proto-oncogenes by placing them under the regulatory elements of the immunoglobulin or T-cell receptor loci.

In many hematopoietic tumors, the translocation allows normally unrelated sequences from two different chromosomes to recombine and form hybrid genes that encode growth promoting chimeric proteins.

Translocation-induced over-expression of a proto-oncogene is best exemplified by Burkitt lymphoma. In Burkitt lymphoma, the most common form of translocation, results in the movement of the c-myc containing segment of chromosome 8 to chromosome 14q band 32.

The molecular mechanisms of the translocation-associated activation of c-myc are variable, as are the precise breakpoints within the gene. In some cases the translocation renders the c-myc gene subject to relentless stimulation by the adjacent enhancer element of the immunoglobulin gene. In others, the translocation causes mutations in the regulatory sequences of the myc gene. In all instances, the coding sequences of the gene remain intact, and the myc gene is constitutively expressed at high levels.

The Philadelphia chromosome characteristic of chronic myeloid leukemia and a subset of acute lymphoblastic leukemias, provides the prototypic example of an oncogene formed by fusion of two separate genes. In these cases, a reciprocal translocation between chromosome 9 and 22 relocates a truncated portion of the proto-oncogene c-abl (from chromosome 9) to the bcr gene encodes for a chimeric protein that has tyrosine kinase activity. Although the translocations are cytogenetically identical in chronic myeloid leukemia and acute lymphoblastic leukemias, they differ at the molecular level. In chronic myeloid leukemia the chimeric protein has a molecular weight of 210KD, whereas in the more aggressive acute leukemias, a slightly different, 180KD, abl-bcr fusion protein is formed.

Normal cell proliferation is controlled by growth factors and cytokines that act on the cell membrane by triggering the cascade of biochemical signals (a process called signal transduction). These signals control the genes that regulate cell growth and division. Oncogenes are altered forms of normal cellular genes called proto-oncogenes that are involved in this cascade of events. They may mutate spontaneously through interaction with viruses, chemicals, or by physical means.

These cellular genes were first discovered by the Noble laureate Michael Bishop and Harold Varmus as passengers within the genome of acute transforming retroviruses, which cause rapid induction of tumors in animals and can also transform animal cells in vitro. Molecular dissection of their genomes revealed the presence of unique transforming sequences not found in the genomes of nontransforming retroviruses. Most surprisingly, molecular hybridization revealed that the viral oncogenes were almost identical to sequences found in the normal cellular DNA.

Most surprisingly, molecular hybridization revealed that the v-onc (viral oncogenes) sequences were almost identical to sequences found in the normal cellular DNA. From this evolved the concept that during evolution, retroviral oncogenes were transuduced (captured) by the virus through a chance recombination with the DNA of a (normal) host cell that had been infected by the virus. Because they were discovered initially as viral genes, proto-oncogenes are named after their viral homologs. Each v-oncogene is designated by the oncogene to the virus from which it was isolated. Thus the v-onc contained in feline sarcoma virus is referred to as v-fes, whereas the oncogene in simian sarcoma virus is called v-sis.

V-oncs are not present in several cancer causing RNA viruses. Example is a group of so called slow transforming viruses that cause leukemia's in rodents after a long latent period. The mechanism by which they cause neoplastic transformation implicates proto-oncogenes. Molecular dissection of the cells transformed by these leukemia viruses has revealed that the proviral DNA is always found to be integrated.

(Inserted) near a proto-oncogene. One consequence of proviral insertion near a proto-oncogene is to induce a structural change in the cellular gene, thus converting it in to a cellular oncogene (c-onc). The strong retroviral promoters inserted in the proto-oncogenes lead to dysregulated expression of the cellular gene. This mode of proto-oncogene activation is called insertional mutagenesis.

Oncogenes are altered growth promoting regulatory genes that governs the cells signal transduction pathways and mutations of these genes leads to either overproduction or increased function of the excitatory proteins. Although oncogenes alone are not sufficient to transform epithelial cells, they appear to be important initiators of the process, and are known to cause cellular changes through mutation of only one gene copy.

Oncogenes or cancer-causing genes are derived from proto-oncogenes. Proto-oncogenes are cellular genes that promote normal growth and differentiation. Proto-oncogenes may become oncogenic by retroviral transduction or by influences that their behaviour in situ, thereby converting them in to cellular Oncogenes.

Several Oncogenes have been implicated in oral carcinogenesis. Aberrant expression of the proto-oncogene epidermal growth factor receptor (EGFR/ c-erb 1), members of the ras gene family, c-myc, int-2, hst-1, PRAD-1, and bcl-1 is believed to contribute towards cancer development. Molecular hybridization of oral carcinogens revealed v-onc (viral oncogene).

Deregulation of growth factors occurs during oral carcinogenesis through increased production and autocrine stimulation. Aberrant expression of transforming growth factor *ά* (TGF-*ά*) is reported to occur early in oral carcinogenesis, first in hyperplasic proliferation by EGFR in an autocrine and paracrine fashion. TGF-*ά* is believed to stimulate angiogenesis and has been reported to be found in normal mucosa in patients who subsequently develop a second primary carcinoma.

Usually oncogenes are dominant as they contain gain-of-function mutations, while mutated tumor suppressors are recessive as they contain loss-of-function mutations. Each cell has two copies of a same gene, one from each parent, and under most cases gain of function mutation in one copy of a particular proto-oncogene is enough to make that gene a true oncogene, while usually loss of function mutation need to happen in both copies of a tumor suppressor gene to render that gene completely non-functional. However, cases exist in which one loss of function copy of a tumor suppressor gene can render the other copy non-functional, and this is called the *dominant negative effect*. This is observed in many p53 mutations.

Rb is a classic example of a tumor suppressor gene. Tumor suppressor genes are normal genes whose loss of function predisposes the cell to cancer. These genes are involved in various important processes critical to cellular homeostasis, including differentiation, DNA repair, control of the cell cycle and apoptosis. Complete loss of function of the retinoblastoma gene is seen in conjunction with a number of tumors including retinoblastomas and melanomas. The relative carcinogenic potency of the various human papillomavirus types is partially explained by the relative avidity of their respective E7 proteins for Rb, *e.g.*, the more tightly a given viral E7 binds Rb, the greater the oncogenic potential of the virus.

When a proto-oncogene is altered to become an oncogene, the pathway of cell growth and proliferation become altered. This may lead to the abnormal growth of cells (neoplastic transformation). More than 100 oncogenes have been identified. An example of an oncogene is the K-ras gene that is mutated in colon cancer cells.

Inheritance pattern of a defective tumor suppresor gene:
Mutations in tumor suppressor genes are, for the most part, loss of function mutations, in which the mutation in the gene results in the absence or

loss of a functional gene product. These mutations are recessive therefore, mutations in both alleles of the gene must occur before a mutant phenotype is observed (figure a). The recessive nature of tumor suppressor mutations allows the heterozygous state to be inherited through the germ line because a single defective allele does not interfere with development of fetus.When one mutant allele of a tumor suppressor gene is inherited, the affected individual does not inherit cancer but inherits a predisposition to develop cancer. If a mutation in the second allele of the tumor suppressor gene occurs, it occurs at the somatic cell level and that single cell will become cancerous. In a normal individual, both alleles of a tumor suppressor gene are active and code for a functioning protein product. Therefore, two separate mutations must occur in a single cell to inactivate both alleles and produce a cancerous cell. In an individual who has inherited one defective allele, every cell in the body carries that defect, and only one mutation at the somatic cell level is necessary to product a totally defective or cancerous cell. Therefore, an individual who inherits one defective allele of a tumor suppressor gene has a great chance of developing cancer than a normal individual and is said to have inherited a predisposition to develop a cancer.

Two–Hit Hypothesis:
Studies on the inheritance patterns of retinoblastoma (RB) by Knudson led to the "two–hit hypothesis" to explain the inherited nature of RB (figure b). According to this hypothesis, a single mutation occurs in the germ line and inactivates one allele of a crucial gene. This mutation is passed on to the off-spring of an affected parent. A second mutation occurs in a somatic cell of the offspring, knocking out the other allele to give a homozygous state and a pre-disposition of the cell to develop cancer. This hypothesis proved correct for RB when it was shown that a single gene at 13q14 was inactivated in families that have the inherited form of RB. This loss of function mutation would be expected to be inherited as an autosomal recessive trait, and in fact this initial mutation does not in itself lead to the development of cancer and is recessive to the normal allele. However, cells with this single allele knocked out have a greater probability of developing cancer if another mutation knocks out the second allele to give a homozygous state. There are several ways in which the second allele can be inactivated. This process, termed loss of heterozygos-ity (LOH), can occur by:

1. Loss of the chromosome with the normal allele or the loss of the chromosome with the normal allele and a duplication of the chromosome with the defective allele.

2. Recombination in the somatic cell by sister chromatid exchange, resulting in the loss of the normal and the presence of two chromosomes, each
   with defective allele.
3. Deletions or point mutations within the normal allele, leading to a loss
   of function and homozygous defective cell.
4. Methylation of the promoter sequences of the normal gene, resulting in
   a loss of transcription and silencing of the second allele.

8.10 Two–Hit Hypothesis

The breast and ovarian cancer susceptibility gene-1 (*BRCA1*) located on chromosome 17q21 encodes a tumor suppressor gene that functions, in part, as a
caretaker gene in preserving chromosomal stability. The caretaker function
of *BRCA1* is a generic one and *BRCA1* mutations confer a specific risk for
tumor types that are hormone-responsive or that hormonal factors contribute
to the etiology, including those of the breast, uterus, cervix, and prostate.

The first breast cancer susceptibility gene, *BRCA1*, was identified and
cloned in 1994 by Miki and colleagues after an intensive search of Miki *et al.*
1994. A year later, a second breast cancer susceptibility gene, BRCA2 was
identified by Wooster et al. 1995, Tavtigian et al. 1996. Mutations of BRCA1
or BRCA2 account for most hereditary breast cancer and breast plus ovarian
cancer families, although there are a sufficient number of non-*BRCA1/BRCA2*
breast cancer families to suggest the existence of at least one additional *BRCA*
gene. Thompson *et al.* 2002 found that *BRCA1* mutation carriers have a significantly increased risk of pancreatic, endometrial, and cervical cancers and
for prostatic cancers in men younger than age 65. *BRCA1* fulfills the criteria
for a tumor suppressor gene (TSG), since the vast majority of cancers that
develop in mutation carriers exhibit loss of the wild-type allele. Venkitaraman
2002, Rosen *et al.* 2003 reported functional roles for BRCA1 in DNA damage
signaling, several different DNA repair processes, apoptosis susceptibility,
and several DNA damage-responsive cell cycle checkpoints.

Aberrant expression of the proto-oncogene epidermal growth factor receptor (EGFR/c-erb 1), members of the ras gene family, c-myc, int-2, hst-1,
PRAD-1, and bcl-1 is believed to contribute towards cancer development.

Oncogenes derived from the c –ras family are often detected in the transfection assay. The family consists of several active genes in both man and rat,
dispersed in the genome. The individual genes, N-ras, H-ras and K-ras are
closely related and code for protein products $\sim$21 kd and known collectively
as $p21^{ras}$.

The H-ras and K-ras genes have v–ras counterparts, carried by the Harvey and Kirsten strains of murine sarcoma virus. Each v-ras gene is closely related to the corresponding c-ras gene, with only a few individual amino acid substitutions. The Harvey and Kirsten virus strains must have originated in independent recombination events in which a progenitor virus gained the corresponding c-ras sequence.

Oncogenic variants of the c-ras genes are found in transforming DNA preparations obtained from various primary tumors and tumor cell lines. Each of the c-ras proto-oncogenes can give rise to a transforming oncogene by a single base mutation. *The mutations in several independent human tumors cause substitution of a single amino acid, most commonly at position 12 or 61, in one of the Ras proteins.*

Position 12 is one of the residues that is mutated in the v–H-ras and v-K-ras genes. *Thus mutations occur at the same positions in v-ras genes in retroviruses and in mutant c-ras genes in human tumors.* This suggests that the normal c-Ras protein can be converted into a tumorigenic form by a mutation in one of a few codons in human.

The general principal established is that *substitution in the coding sequence can convert a cellular proto-oncogene into an oncogene.*

The ras genes appear to be finely balanced at the edge of oncogenesis. Almost any mutation at either position 12 or 61 can convert a c-ras proto-oncogene into an active oncogene.

When the expression of a normal c-ras gene is increased, either by placing it under control of a more active promoter or by introducing multiple copies into transfected cells, recipient cells are transformed. Some mutant c-ras genes that have changes in the protein sequence also have a mutation in an intron that increases the level of expression (by increasing processing of m RNA ~10x). Also, some tumor lines have amplified ras genes. A 20-fold increase in the level of a nontransforming Ras protein is sufficient to allow the transformation of some cells. The effect has not been fully quantitated, but it suggests the general conclusion that oncogenesis depends on over –activity of Ras protein, and is caused either by increasing the amount of protein or (more efficiently) by mutations that increase the activity of the protein.

Ras is a monomeric guanine nucleotide–binding protein that is active when bound to GTP and inactive when bound to GDP. It has an intrinsic GTPase activity.

Gene amplification and deregulation of the expression of the *c-MYC* oncogene located in region 8q24 region are the main activating mechanisms of this oncogene in human tumors. Its protein (myc) enables the cell to enter the cell cycle, activating several genes including those encoding ornithine decarboxylase, phosphatase cdc25A and the transcription factor E2F (Zajac-Kaye, 2001; Mai et al., 2003) and repressing others (such as the p27 releasing protein from the cdk2 complex) by means of the activity of the cyclin E-cdk2 complex. Over expression of the c-myc protein can increase cell proliferation, impair differentiation and lead to an increase in cell apoptosis (Abba et al., 2004), although expression of this protein usually decreases when the cell is leaving the cell cycle and during differentiation. Its over expression is a major factor that can be used to distinguish between well differentiated adenomas and adenocarcinomas.

Deregulation of growth factors occurs during oral carcinogenesis through increased production and autocrine stimulation. Aberrant expression of transforming growth factor α(TGF-α) is reported to occur early in oral carcinogenesis, first in hyperplastic epithelium, and later in the carcinoma within the inflammatory cell infiltrate, especially the eosinophils, surrounding the infiltrating epithelium. TGF-α stimulates cell proliferation by binding to EGFR in an autocrine and paracrine fashion. TGF-α is believed to stimulate angiogenesis and has been reported to be found in "normal" oral mucosa in patients who subsequently develop a second primary carcinoma.

EGFR, the biological receptor of EGF and TGF-α is frequently over expressed in oral cancers, and this was found to be the result of EGFR gene amplification in 30% of oral cancers. It has been suggested that over expression of the EGF receptor is often accompanied by the production of its ligands, TGF-α and EGF. The interaction of the receptor and its ligands initiates a cascade of events, translating extra cellular signals through the cell membrane and triggering intrinsic tyrosine kinase activity. Mutations of genes encoding growth factor receptors can result in an increased number of receptors, or the production of a continuous ligand independent mitogenic signal. Gene amplification and increased numbers of EGF receptors in oral cancers are associated with the degree of differentiation and aggressiveness of the tumors. Oral squamous carcinomas over expressing EGFR have been shown to exhibit a greater response to chemotherapy, compared with EGFR negative tumors. This is presumably because of the higher proliferative activity in the tumors over expressing EGFR, leading to a higher sensitivity to cytotoxic drugs, and current data suggest that the therapeutic application of antibodies directed against EGF receptors might be useful in the treatment of premalignant and malignant lesions. During oral carcinogenesis, intracellular messengers might also be intrinsically activated, thereby overriding the necessity for ligand–receptor regulated signals. Among the genes involved in intracellular signaling pathways, members of the ras family have been examined in oral cancers. The ras oncogenes represent the best example of activation by point mutation. A large number of human tumors carry ras mutation. The frequency of such mutation varies with different tumors, but in some types it is high. H-ras, K-ras, and N-ras all encode the protein p21, which is located on the cytoplasmic membrane of the cell membrane, and which transmits mitogenic signals by binding GTP. The mitogenic signal is terminated by the conversion of GTP to GDP by hydrolysis, but when the ras oncogene is mutated this conversion can be prevented, thus leading to continuous stimulation.

Some studies have indicated that members of the ras oncogene family are over expressed in oral cancers. Although loss of control of N-ras might be an early step in carcinogenesis in oral cancers, with increased expression occurring early in dysplastic lesions, ras mutations are uncommon in the progression of oral cancers in the Western world, occurring in less than 5% of all cases. In contrast, 55% of lip cancers have H-ras mutation and H-ras mutation occurs in 35% of oral cancers in the Asian population, where it is especially associated with betel nut chewing.

Transcription factors that activate other genes are also activated in oral cancer. The functional activity of many of these proteins is regulated by receptor activated second messenger pathways, and neutralization of these genes could result in a cell cycle block, preventing mitogenic and differentiation responses to growth factors. Among these genes c-myc, which helps regulate cell proliferation, is frequently over expressed in oral cancers as a result of gene amplification. In 1999 a unpublished work by HK Williams *et al.* stated that Over expression is frequently associated with poorly differentiated tumors, although more recently c-myc has been shown to be over expressed in moderate and well differentiated oral carcinomas, in which cell proliferation far outweighed the number of apoptotic cells present. c-Myc induces both cell proliferation and apoptosis. c-Myc requires p53 to induce apoptosis and the retinoblastoma tumor suppressor gene Rb-1 nuclear protein pR6 interacts with the c-myc gene, preventing its transcription, and thus inhibiting cell proliferation. However, on phosphorylation of pR6, c-Myc is increased and cell proliferation proceeds. In this regard, it is interesting found that pR6, c-myc, and p53 to be expressed in all oral carcinomas, irrespective of differentiation. However, further studies are needed to determine the genetic status of these oncogenes and tumor suppressor genes, and to measure the concentrations of the proteins, to determine possible controlling pathways in the development of these oral cancers.

The PRAD-1 gene located on 11q13 encodes cyclin D, which together with the Rb gene product controls the G1 to S transition of the cell cycle. The PRAD-1 gene is amplified in 30–50% of head and neck cancers. Amplification of PRAD-1 is correlated with cytological grade, infiltrative growth pattern, and metastases.

The hst-1/int-2 gene encodes a protein that is homologous to fibroblast growth factor, and which in oral cancers has been shown to be involved in tumor growth, and to have angiogenic activity. This gene maps to human chromosome 11q13.3, which is co amplified with int-2 in some cancers. Lese *et al.*, in 1995, reported co amplification of the int-2 and hst-1 genes in oral squamous carcinomas, and other authors have suggested that this co amplification in head and neck squamous carcinomas is associated with tumor recurrence and progression of disease.

Int-2 amplification in premalignant lesions adjacent to neoplasia, both in areas of dysplasia and hyperplasia, which suggests that int-2 can be amplified before tumor development.

8.11 p53 (Guardian of Genome)

p53 is a nucleo phosphoprotein which act as tumor suppressor comprising 393amino acids, and was discovered in 1970. It is located on chromosome 17p13.1 and is the single most common target for genetic alteration in human tumors. As with Rb gene, inheritance of one mutant p53 allele predisposes individuals to develop malignant tumors including carcinomas, sarcomas, lymphomas and brain tumors. The physiologic function of the p53 protein is that of preventing accumulation of genetic damage in cells either by allowing for repair of the damage before cell division or by causing death of cell. The normal p53 has a very short half life, the quantity in normal cells is extremely small therefore it is not usually detectable by Immunohistochemistry. Mutant p53 has a prolonged half life and can accumulate in cells to levels that are detectable.This mutant protein is normally not active, thus leading to the loss of tumor suppressor function of protein. In most instances mutation that inactivate both copies of the p53 gene are acquired in somatic cells. Less commonly some individual inherit a mutant p53 allele. *A cell with damaged DNA that cannot be repaired is directed by the p53 gene to undergo apoptosis.* So it has been rightfully called as "Guardian of the genome". Homozyous loss of p53, DNA damage goes unrepaired, mutations become fixed in dividing cells and the cells reluctantly turns onto a malignant transformation.

Activation of normal p53 by DNA–damaging agents or by hypoxia leads to cell–cycle arrest in G1 and induction of DNA repair, by transcriptional up–regulation of the cyclin–dependent kinase inhibitor p21 and GADD45 genes. Successful repair of DNA allows cells to proceed with cell cycle; if DNA repair fails, p53 – induced activation of the BAX gene promotes apoptosis. In cells with loss or mutations of p53, DNA damage does not induce cell cycle arrest or DNA repair, and hence genetically damaged cells proliferate, giving rise eventually to malignant neoplasms.

p53 have at least three important function involved in its tumor suppressor role.

- In response to DNA damage, p53 stops cell division
- Up regulates genes involved in DNA repair such as GADD45
- If the DNA cannot be repaired, p53 induce programmed cell death.

This ensures that no unrepaired DNA damage is propagated and is vitally important to maintaining the integrity of the genome because of these crucial function p53 has been termed as "guardian of the genome".

8.12 Guardian of Genome

It plays an important role in control of the cell cycle, acting as a factor in transcription, genomic stability, cell differentiation and apoptosis. Aberrations of the p53 gene are the most common genetic alterations in oral cancer. Detection of this protein usually indicates that stabilizing mechanisms are inefficient, that is, there is a loss of pro-apoptotic function, giving rise to continued tumor growth. This gene is not detected in the immunohistochemical study of normal cells. The detection of p53 in pre-invasive adjacent areas in squamous carcinoma and dysplastic lesions suggests that it may constitute an advance in the natural history of oral cancer. Various studies have shown that expression of the p53 protein in biopsies where there are *in situ* oral dysplasias and carcinomas is preceded, in a period of months or weeks, by malignant histological changes. However, it cannot be concluded that it is an intermediate biomarker of risk as its mutation appears relatively late in the carcinogenic process—despite the fact that, as Bautista and Santiago report, the immuno localization of p53 appears in the very early stages of squamous cell carcinoma. Whatever the case, the mutation of p53 and/or its over expression are themselves not sufficient for the development of oral carcinoma. This alteration appears in between 11 and 80% of aerodigestive carcinomas. A recent study by (Schildt et al.) found p53 to be over expressed in 63%

of oral carcinomas, with p53 mutations in 36%. Altered p53 expression in premalignant lesions is associated with increased chromosomal polysomy.

Tumor suppressor genes code for antiproliferation signals and proteins that suppress mitosis and cell growth. Generally tumor suppressors are transcription factors that are activated by cellular stress or DNA damage. Often DNA damage will cause the presence of free floating genetic material as well as other signs, and will trigger enzymes and pathways which lead to the activation of tumor suppressor genes.

The functions of such genes is to arrest the progression of cell cycle in order to carry out DNA repair, preventing mutations from passed on to daughter cells. The tumor suppressor genes are most often inactivated by point mutations, deletions and rearrangements in both gene copies. However, a mutation can damage the tumor suppressor gene itself. The invariable consequence of this is that DNA repair is hindered or inhibited. DNA damage accumulates without repair, inevitably leading to cancer.

There has been much research on the tumor suppressor gene p53. The p53 protein blocks cell division at the G1 to S boundary, stimulates DNA repair after DNA damage, and also induces apoptosis. These functions are achieved by the ability of p53 to modulate the expression of several genes. The p53 protein transcriptionally activates the production of the p21 protein, encoded by the WAF1/CIP gene, p21 being an inhibitor of cyclin and cyclin dependant kinase complexes. p21 transcription is activated by wild-type p53 but not mutant p53. However, WAF1/CIP expression is also induced by p53 independent pathways such as growth factors, including platelet derived growth factor (PDGF), fibroblast growth factor (FGF), and transforming growth factor ß(TGF-ß). Wild-type p53 has a very short half life (four to five minutes), whereas mutant forms of protein are more stable, with a six hour half life.

Mutation of p53 occurs either as a point mutation, which results in a structurally altered protein that sequesters the wild-type protein, thereby inactivating its suppressor activity, or by deletion, which leads to a reduction or loss of p53 expression and protein function.

The tumor suppressor gene p53 is known to be mutated in approximately 70% of adult solid tumors.

p53 has been shown to be functionally inactivated in oral tumors, and restoration of p53 in oral cancer lines and tumors induced in animal models has been shown to reverse the malignant phenotype. Smoking and tobacco use have been associated with the mutation of p53 in head and neck cancers.

Other tumor suppressor genes include doc-1, the retinoblastoma gene, and APC. The doc-1 gene is mutated in malignant oral keratinocytes, leading to a reduction of expression and protein function. The precise function of the Doc-1 protein in oral carcinogenisis is unclear, but it is very similar to a gene product induced in mouse fibroblasts by tumor necrosis factor α(TNF-α). Normally, TNF-α decreases proliferation and increases differentiation, and has been shown in oral squamous cell carcinoma cell lines to be responsible, either alone or in combination with interferons α or γ, for antiproliferative activity.

More than 50% of all human oral cancer cells are defective in p53. This finding has led to a new approach to treat cancer by directly killing the tumor cells that are defective in p53. The procedure relies on the use of defective adenoviruses that cannot replicate in cells expressing the tumor suppressor protein p53. Normal adenoviruses contain a gene that codes for a protein called E1B. During the infection process, the E1B protein binds to the p53 protein and stimulates the cells to enter the S phase of the cell cycle and support DNA replication of the virus. Thus, viruses have evolved proteins like E1B to block the function of p53 and stimulate cells to bypass the G1 block and enter the S phase of the cell cycle. An adenovirus mutant that lacks the E1B protein should be unable to replicate in cells that express the p53 protein. In the presence of a functioning p53, the virally infected cells are blocked in the G1 phase of the cell cycle and will not enter into the S phase; thus the virus is unable to replicate its DNA and the infection is aborted. However a cell lacking the p53 protein should serve as a receptive host for the defective adenovirus because the cell is no longer blocked in G1 in response to the viral DNA thus it progresses to the S phase and allows the defective virus to replicate and kill the host cells. Because over 50% of human cancer cells are defective in p53, a mutant adenovirus that does not express E1B could potentially be used to target these cancer cells and specifically kill them. Any normal cells present should use their p53 protein to prevent the replication of the defective virus and should remain viable. The therapy currently is limited to tumors that can receive direct injections, but even if it works on a few cancers it will be an important and beneficial treatment.

The most important tumor suppressor gene is p53 discovered in 1977 by Sarly. p53 is a nuclear phosphoprotein and was originally discovered in SV-40 transformed alleles where it is associated with T-antigen.

P53, is located on chromosome 17q13.1 and is the single most common target for genetic alteration in human tumors. Homozygous loss of p53 gene is found in virtually every type of cancer. Mutations that inactivate both the copies of p53 alleles are acquired in somatic cells or inherited to Li-Fraumeni syndrome. Under physiological conditions it has short half life (20 mines) and does not police the normal cell cycle.

p53 is called to apply emergency brakes when DNA is damaged by irradiation, UV light or mutagenic chemicals.

As with Rb gene, inheritance of one mutant p53 allele predisposes individuals to develop malignant tumors including carcinomas, sarcomas, lymphomas and brain tumors.

There is a rapid increase in p53 levels and activation of p53 as a transcription factor. The accumulated wild type p53 binds to DNA and stimulates transcription of several genes that mediate the two major effects of p53: cell cycle arrest and apoptosis.

p53 induced cell cycle arrest occurs late in the G1 phase and is caused by the p53 dependent transcription of the CDK inhibitor p21. The p21 gene, inhibits the cyclin/CDK complexes and prevents the phosphorylation of pRb necessary for cells to enter the S phase.

A pause in the cell cycle brought about by the action of p53 is welcome because it allow the cells time to repair DNA damage inflicted by mutagens. p53 also helps this process by inducing transcription of some DNA repair enzymes.

The fact that p53 mutations are common in a variety of human tumors suggests that the p53 protein serves as a gatekeeper against the formation of cancer. Indeed it is evident that p53 act as a 'molecular policeman' that prevents the propagation of genetically damaged cells.

The ability of p53 to control apoptosis in response to DNA damage has some practical therapeutic implications. Radiation and chemotherapy, the two common modalities of cancer treatment, mediate their effects by including DNA damage and subsequent apoptosis.

Mutation of the p53 tumor suppressor gene may represent the most common genetic change in human cancer. The physiologic function of the p53 protein is that of preventing accumulation of genetic damage in cells either by allowing for repair of the damage before cell division or by causing death of the cell. The normal p53 protein has a very short half-life; the quantity in normal cells is extremely small. Therefore, it is usually not detectable by immunohistochemistry. Mutant p53 protein has a prolonged half-life and can accumulate in cells to levels that are detectable. This mutant protein is normally not active, thus leading to the loss of the tumor suppressor function of the protein. More than 50% of oral squamous cell carcinomas are positive for p53 protein, and mutations of the p53 gene have been documented.

p53 senses DNA damage and assists in DNA repair by causing GI arrest and inducing DNA repair genes. A cell with damaged DNA that cannot be repaired is directed by p53 to undergo apoptosis. In views of these activities, p53 has been rightfully called a "guardian of the genome".

A recent immunohistochemical study of p53 and p21, one of the downstream target genes activated by p53, in 53 oral verrucous leukoplakias reported that aberrant immunoreactivity of p53 and p21 was closely associated with malignant transformation. It has been reported earlier that, in most oral squamous cell carcinomas, p21 expression does not depend on p53 status, whereas in another study, the p21 expression seemed to correlate with p53 status.

In addition to somatic and inherited mutations, p53 gene functions can be inactivated by other mechanisms. pRb, the transforming proteins of several DNA viruses, including the E6 protein of human papillomaviruses, can bind to the degree of p53. The cellular p53 binding protein, mdm2, which normally down regulates p53 activity, is over expressed in a subset of human soft tissue sarcomas as a result of gene amplification. By promoting rapid degradation of p53, mdm2 acts as an oncogene.

The important tumor suppressor gene involved in viral carcinogenesis is p53, which has at least three important functions involved in its tumor suppressor role (see Figure 14). In response to DNA damage, p53 stops cell division and up-regulates genes involved in DNA repair such as Gadd45. If the DNA cannot be repaired, p53 performs its third and perhaps most crucial function, to induce programmed cell death. This ensures that no unrepaired DNA damage is propagated, and is vitally important to maintaining the integrity of the genome. Because of these crucial functions, p53 has been termed the "guardian of the genome". Therefore, loss of p53 promotes genetic instability and strongly predisposes affected cells to accumulate additional genetic abnormalities.

Because p53 is also important in regulating differentiation and suppressing cell division, human papillomaviruses have evolved the E6 protein to circumvent it. Analogous to E7 proteins, E6 proteins of the high risk viruses bind p53 with greater avidity. In addition, the most oncogenic viruses actually promote ubiquitin-mediated p53 breakdown, leading to a profound loss of p53 activity.

p73 found in late 1997 is dubbed the big brother of p53. Located on 1p36, this gene encodes a protein that bears many similarities to p53. It has a DNA binding domain that resembles the corresponding region of p53, and similar to the later it can cause cell cycle arrest as well as apoptosis under appropriate conditions. Deletions of 1p36, where the p73 gene resides, are common in a variety of tumors, including neuroblastomas and colon and breast cancers.

BRCA-1, on chromosome 17q12-21, and BRCA-2, on chromosome 13q12-13, are two recently discovered tumor suppressor genes that are associated with the occurrence of breast and several other cancers. Protein products of both genes are localized to the nucleus and are involved in transcriptional regulation. The BRCA-1 and BRCA-2 proteins interact with Rad 51, a protein implicated in the regulation of recombination and double stranded DNA repair.

Mutations in BRCA genes, similar to mutations in other DNA repair genes, do not directly regulate cell growth; rather they predispose to errors in DNA replication, thus leading to mutations in other genes that directly affect cell cycle and cell growth.

Several types of molecules expressed on the cell surface can regulate cell growth and behavior. Such molecules include receptors for growth inhibitory factors, such as TGF-ß, and proteins that regulate cellular adhesion, such as the cadherins. The binding of TGF-ß to its receptors up-regulates transcription of growth inhibitory genes.

It mediates this effect, in part, by stimulating the synthesis of cyclin dependent kinase (CDK) inhibitors. These block the cell cycle by inhibiting the actions of cyclin/CDK complexes. Mutations of the TGF-ß receptor and its signaling pathway have been discovered in many cancers.

Cadherins are a family of glycoproteins that acts as glues between epithelial cells. Loss of cadherins can favor the malignant phenotype by allowing easy desegregations of cells, which can then invade locally or metastasize.

Deleted in colon carcinoma (DCC) is a gene located on chromosome 18q21. Because this chromosome region is frequently deleted in human colon and rectum carcinomas, the DCC gene has been considered a candidate tumor suppressor gene. Its structure resembles other cell surface molecules that are involved in cell to cell and cell to matrix interactions; hence it was proposed that the DCC gene may regulate cell growth and differentiation by integrating signals from the cell's environment. Thus it seems that some other gene in close linkage with DCC on chromosome 18q21 is the real culprit for carcinogenesis.

8.12.1 p16^{INK4a} and p14^{ARF}

A number of genetic aberrations have been reported in end-stage squamous cell carcinoma of the head and neck, including p16^{INK4a} and p14^{ARF} (INK4a/ARF). Still, the cell cycle-regulatory genes p16^{INK4a} and p14^{ARF} remain poorly understood in oral cavity premalignant lesions.

The Ink4a/Arf locus encode two tumor suppressor proteins, p16^{Ink4a} and p14^{ARF} regulate the activities of the Rb^4 protein and the p53 transcription factor, respectively. The p16^{Ink4a} protein inhibits the activities of cyclin D-dependent kinases, Cdk4 and Cdk6, to prevent their ability to phosphorylate and inactivate Rb and other Rb-family proteins, thereby arresting cells in the G$_1$ phase of the cell cycle. By contrast, p14^{Arf} blocks various activities of the p53-negative regulator Mdm2, leading to p53 stabilization and a p53-dependent transcriptional response.

The atypical structure of the INK4a/ARF locus in 9p21 encodes unrelated tumor suppressor proteins, p16^{INK4a} and p14^{ARF} (the human counterpart of murine p19^{ARF}). These are specified by different first exons that are spliced to a common second exon translated in alternative reading frames, their expression being controlled by independent promoters. p16^{INK4a}, a specific inhibitor of cyclin D-dependent kinases, contributes to G1 arrest by blocking Rb phosphorylation. On the other hand, p14^{ARF} interferes with all of the known functions of Hdm2 (eg, direct inhibition of p53-mediated transactivation, ubiquitin ligase activity, and nuclear export of p53), possibly through induction of Hdm2 degradation, indirectly leading to an increase in the activity and stability of p53. p14^{ARF} is induced by inappropriate hyperproliferative signals (such as *myc*, E2F-1, *ras*, E1A, v-*abl*) and mediates p53 activation in response to oncogenic stimuli. Specifically, responsiveness of the p14^{ARF} promoter to E2F-1 makes p14^{ARF} a nexus between the Rb and p53 pathways. p53 suppresses p14^{ARF} expression through a poorly understood mechanism, which generates an additional regulatory circuitry.

p14^{ARF} is a highly basic protein that localizes to the nucleolus. When induced, p14^{ARF} binds to Hdm2, thereby allowing p53 to stabilize and accumulate in the nucleoplasm. Classically, this p14^{ARF}-Hdm2 binding has been assumed to take place in the nucleolus, although antagonization of Hdm2 by p14^{ARF} independently of nucleolar localization has recently been reported. In addition, it has been suggested that p14^{ARF}-p53 direct binding without requirement for Hdm2 as a bridging molecule is also possible, although the functional implications of this interaction remain unknown.

Both human and murine p14^{ARF} contact the central acidic domain of Hdm2 through independent binding sites. Additionally, nucleolar localization sequences (NrLS) have been mapped in exons 1β and 2 of p14^{ARF}. These motifs are required for the localization of p14^{ARF} to the nucleolus; mutations in the p14^{ARF} NrLS have been described as impeding the correct localization of this protein, resulting in its nucleoplasmic accumulation and the consequent loss of its ability to stabilize p53.

In human tumors, the p53 gene is inactivated by mutation in more than 50% of cases; in a high proportion of the rest, the p53 pathway would be expected to be disrupted by *Hdm2* amplification or p14^{ARF} loss. In some cancers, the frequency of p14^{ARF} alteration is remarkably high; deletions affecting the 9p21 region (eg, in the cases of glioblastoma and astrocytoma) and hypermethylation of CpG islands in the p14^{ARF} promoter (eg, in the case of gastric cancer) are the main inactivation mechanisms. Point mutations are infrequent and usually also affect p16^{INK4a}. In other neoplasias, p14^{ARF} loss appears to be a rarer event.

The high frequency of mutation, deletion, and promoter silencing of the gene encoding p16^{INK4A} (p16) in premalignant dysplasias and squamous cell carcinomas (SCC) of epidermis and oral epithelium classifies p16 as a tumor suppressor. However, the point during neoplastic progression at which this protein is expressed and presumably impedes formation of an SCC is unknown. Induction of p16 has been found to be responsible for the senescence arrest of normal human keratinocytes in culture, suggesting the possibility that excessive or spatially abnormal cell growth *in vivo* triggers p16 expression.

Molecular alterations in a number of cell cycle-regulatory genes have been identified in end-stage SCC of the head and neck, including INK4a/ARF inactivation rates of 70–85%. The INK4a/ARF locus is located on chromosome 9p21 and has the unique distinction of encoding two cell cycle-regulatory genes, p16^{INK4a} and p14^{ARF}. Briefly, alternative splicing of the first exon and common downstream exons permits one gene to encode two different products, which function via two distinct pathways to inhibit cell cycle progression. The tumor-suppressive activity of p16^{INK4a} is described to its ability to bind both cdk4 and cdk6. This in turn inhibits the catalytic activity of the cdk4/6-cyclin D complex, blocks retinoblastoma phosphorylation, and ultimately prevents cell cycle progression. In contrast, p14^{ARF} interacts with the oncogenic protein MDM2, inducing stabilization of p53 and enhancing p53-related functions. A single alteration in the INK4a/ARF locus can potentially disrupt the p16–Rb and p14–p53 tumor suppressor pathways and facilitate cancer development.

Numerous studies recognize the prominent tumor suppressor function of p16^{INK4a}; however, INK4a/ARF locus alterations in premalignant oral disease remain incompletely investigated and poorly understood. The main modes of p16^{INK4a} inactivation in SCC of the head and neck are known to include homozygous deletions, mutations, and gene hypermethylation events.

APC controls certain transcriptional activities because loss of APC function in colon carcinoma cells results in cytoplasmic ß-catenin transducing Wnt signals by associating with T cell factor (TCF) and lymphoid enhancer factor (LEF). These complexes then pass to the nucleus where they activate Tcf target genes in an uncontrolled manner, which might contribute to colon tumorigenesis. Although few cases of mutant APC have been described in oral carcinoma, further investigations are required to investigate the function of ß-catenin in oral cancer.

Recently, it has been shown that APC might indirectly regulate the E-cadherin–catenin complex because in E-cadherin negative colon carcinoma cell lines, ß-catenin is preferentially bound to APC. If, however, these cell lines are transected with E-cadherin, ß-catenin redistributes from the APC bound complex to the E-cadherin–catenin complex and is accompanied by growth inhibition and decreased tumorigenicity. At present, however, it is unknown whether APC controls the E-cadherin–catenin complex in oral carcinoma.

Other cell adhesion molecules include the action dependent, heterodimeric family of integrins, which mediate cell–cell and cell–matrix interactions, and play a role in the maintenance of tissue integrity and in the regulation of cell proliferation, growth, differentiation, and migration. Integrin expression is variable between different tumor types but these molecules are implicated in tumor progression and metastasis.

In oral squamous cell carcinomas, there is variable loss or reduced expression of ß1 integrins and α6ß4, especially in poorly differentiated carcinomas. The localization and the quantity of the α6 chain has been shown to be altered, with high levels of α6 in contrast to ß4, both in premalignant and malignant oral mucosa. This suggests that this might be an early but non-specific marker of oral malignancy, and that abnormal extra cellular signals might be involved. Recently, it has been reported that metastatic oral squamous carcinoma cell lines show strong expression of α2–6 integrins compared with non-metastatic cell lines, and the pronounced expression seen in primary oral squamous carcinomas correlates significantly with the mode of tumor invasion and nodal involvement.

The expression of αv is also altered, with $\alpha v\beta 6$ being expressed in malignant oral carcinomas. This integrin is not expressed in normal epithelium, suggesting that it might play a role in tumor migration. Reduced expression is seen for $\alpha v\beta 5$ in oral cancers in contrast to normal epithelium. This integrin might be important in oral neoplasia because in vitro studies suggest that αv negative malignant cell lines can be reversed after transfection of the integrin.

Squamous cell carcinomas of the oral cavity are characterized by their ability to spread locally and regionally, this being associated with a high rate of fatality, with a breach in the basement membrane separating epithelial and mesenchymal compartments being the first step in tumor invasion. Not only does the alteration in expression and/or function of several of these cell adhesion molecules result in tumor infiltration and metastasis, but there is now compelling evidence that urokinase 92 kDa type IV collagenase (MMP-9) is produced by tumor cells, whereas type I collagenase and stromeolysins 2 and 3 are synthesized by stromal cells. However, 72 kDa type IV collagenase (MMP-2), also found in oral cancers, is stimulated by MT-MMP, a cell membrane metalloproteinase. Recent evidence suggests that stromal cells influence the synthesis of urokinase and MMP-9, although the transcriptional requirements and cell signaling pathways remain to be fully elucidated.

A novel inhibitor of apoptosis, survivin, plays a role in oncogenesis survivin plays an important role during oral carcinogenesis, and that the gene expression may be regulated by an epigenetic mechanism.

8.12.2 Cyclin-dependent kinase inhibitors

There are two families of CDKIs: the p21 family and the INK4 family. p21 is the universal inhibitory gene of the CDKs, and is localized on chromosome6. Under normal conditions it forms a complex with cyclins. An association has been found between p21 expression and the degree of tumor differentiation. It is likely that the over expression of p21 is caused by p53-independent transactivation mechanisms.

The p21 protein functions by binding to multiple cyclin–Cdk complexes and blocking their kinase activity, thus blocking the phosphorylation of proteins needed for progression in the cell cycle. The binding of p21 to the G1 cyclin–Cdk complexes, Cdk4–cyclin D and Cdk2–Cyclin E, is central to the G1 arrest that follows DNA damage by ionizing radiation. An arrest of the cell in G1 allows time for the DNA repair systems to correct the damage.

Another function of p21 is to bind proliferating cell nuclear antigen (PCNA), which is a factor required to activate DNA polymerase. With p21 bound to PCNA, DNA replication is inhibited, thus preventing the damage from being replicated before it is repaired. Thus, the ability of a cell to stop progression in the cell cycle to allow repair of damaged DNA to occur is an important control process that prevents the occurrence of mutations that might lead to a cancerous cell.

Various phases of cell cycle are orchestrated by cyclins and cyclin dependent kinases (CDKs) and their inhibitors. Mutations in genes that encode this cell cycle have been found in several human cancers.

Cyclin dependent kinases are expressed constitutively during the cell cycle but in an inactive form. They are activated by phosphorylation after binding to another family of proteins, called cyclins. Cyclins are synthesized during specific phases of the cell cycle, and their function is to activate the CDKs on completion of this task, cyclin levels decline rapidly. Each phase of the cell cycle circuitry is carefully monitored, the transition from G1 to S is an extremely important checkpoint in the cell cycle clock because once cells cross this barrier they are committed progress in to S phase. When a cell receives growth promoting signals, the synthesis of D type cyclins that bind CDK4 and CDK6 is stimulated in the early part of G1. Later in the G1 phase of the cell cycle, the synthesis of E cyclin is stimulated, which in turn, binds to CDK2. The cyclin D/CDK4, CDK6 and cyclin E/CDK2 complexes phosphorylated with the retinoblastoma proteins (pRb). pRb binds to the E2F family of transcription factors. Phosphorylation of pRb unshackles the E2F proteins and they in turn activate the transcription of several genes whose products are essential for progression through the S phase. These include DNA polymerases, thymidine kinase, and dihydrofolate reductase. The progress of cells from the S phase in to the G2 phase is facilitated by up-regulation of cyclin A which binds to CDK2 and to CDK1.

Early in the G2 phase, B cyclin takes over. By forming complexes with CDK1, it helps the cell move from G2 to M.

The activity of CDKs is regulated by two families of CDK inhibitors (CDKIs). One family of CDKIs composed of three proteins, called p21, p27 and p57, inhibits the CDKs broadly. Whereas the other family of CDKIs has selective effects on cyclin D/CDK4 and cyclin D/CDK6.

The four members of this family (p15, p16, p18, p19) are sometimes called INK4 proteins (because they are inhibitors of CDK4 and CDK6).

Amplification of CDK4 gene occurs in melanomas, sarcomas, and glioblastomas. Mutations affecting cyclin B and cyclin E and other CDKs also occur in certain malignant neoplasms but they are much less frequent than those affecting cyclin D/CDK4.

8.12.3 Retinoblastoma protein (pRb)

The retinoblastoma (RB) gene spans over 200 kb of DNA and codes for a protein (pRb) with molecular weight of 105 to 110 Kd. The Rb gene product (pRb) is a DNA – binding protein hence it is a key player in the regulation of cell cycle. pRb is expressed in the nucleus of all cells; it regulates several important cell cycle progression, cell differentiation, and apoptosis and is known to act at the level of gene transcription by forming protein–protein complexes with both upstream TFs and basal TFs. This interaction can result in either positive or negative stimulation of expression of the genes coding for proteins that regulate progression through the cell cycle. The central role of pRb in regulating the cell cycle is attested to by the discovery of a variety of growth promoting and growth inhibiting pathways all of which coverage of pRb (figure on next page). It is expressed in every cell type examined where it exist " in an active unphosphorylated and an inactive phosphorylated state". Quiescent cells contain the active unphosphorylated pRb which prevents DNA sythesis. When the cells in G1 are stimulated by growth factor, the concentration of D cyclins goes up and resultant activation of the cyclin D/CDKs leads to phosphorylation of pRb. Phosphorylation of pRb sets free the transcription factors which trigger DNA synthesis.

This protein is a key factor in the G1 check point and is therefore the key to the R point. Koontongkaew et al. found this protein to be 58.49% over expressed in the oral carcinomas they studied. Deregulation of the pRb gives rise to aberrations in various cell proteins such as CD1 and CDK4; this mechanism is necessary for the development of oral and pharyngeal cancer.

The signal transducing proteins are oncoproteins that mimic the function of normal cytoplasmic signal transducing proteins.

Such proteins are strategically located on the inner leaflet of the plasma membrane, where they receive signals from outside the cell by activating GFR and transmit them to the cells nucleus. The best and most well studied example of a signal transducing oncoprotein is ras family of guanine triphosphate (GTP) binding proteins.

Most such proteins are strategically located on the inner leaflet of the plasma membrane, where they receive signals from outside the cells (e.g. by activation of growth factor receptors) and transmit them to the cell nucleus. The signal transducing proteins are heterogeneous. The best example of a signal transducing oncoprotein is the ras family of guanine triphosphate (GTP)-binding protein.

The ras proteins were discovered initially in the form of viral oncogenes. Approximately 10 to 20% of all human tumors contain mutated versions of ras proteins.

Mutation of the gene is the single most common abnormality of dominant oncogenes in human tumors. ras plays an important role in mutagenesis induced by growth factors for example, blockade of ras function by microinjection of specific antibodies blocks the proliferative response to EGF, PDGF, and CSF-1.

In the inactive state, ras proteins bind guanosine diphosphate (GDP). When cells are stimulated by growth factors or other receptor ligand interactions, ras becomes activated by exchanging GDP for GTP. The activated ras excites the MAP kinase pathway by recruiting the cytosolic protein raf-1. The MAP kinases activated target nuclear transcription factors and thus promote mitogenesis.

In normal cells, the activated signal transmitting stage of ras protein is transient because its intrinsic GTPase activity hydrolyzes GTP to GDP, thereby returning ras to its quiescent ground state.

The orderly cycling of the ras protein depends on two reactions:

1. Nucleotide exchange (GDP by GTP) which activates ras protein, and
2. GTP hydrolysis, which converts the GTP-bound inactive form.

The removal of GDP and its replacement by GTP during ras activation is catalyzed by a family of guanine nucleotide releasing proteins that are recruited to the cytosolic aspect of activated growth factor receptors by adaptor proteins. The GTPase activity is accelerated by GTPase activating proteins (GAPs). These widely distributed proteins bind to the active ras and augment its GTPase activity by more than 1000-fold leading to rapid hydrolysis of GTP to GDP and termination of signal transduction. Thus, GAPs function as 'brakes' that prevent uncontrolled ras activity.

Mutant ras proteins bind GAP, but their GTPase activity fails to be augmented. Hence the mutant proteins are 'trapped' in their excited GTP- bound form, causing in turn, a pathologic activation of the mitogenic signaling pathway.

ras is also involved in the regulation of cell cycle, as the passage of cells from G0 to the S phase is modulated by a series of proteins called cyclins and cyclin dependent kinases (CDKs). ras controls the levels of CDKs.

To block ras activity, ras must be anchored under the cell membrane close to the cytoplasmic domain of the growth factor receptors. Such anchoring is made possible by attachment of an isoprenyl lipid group to the ras molecule by the enzyme farnesyl transferase. The farnesyl moiety forms the bridge between ras and the lipid membrane. Inhibitors of farnesyl transferase can disable ras by preventing its normal localization.

In addition to ras, several non-receptor associated tyrosine kinases also function in the signal transduction pathways. The mutant forms of non-receptor associated tyrosine kinases are commonly found in the forms of v-oncs in animal retroviruses (e.g. v-abl, v-src, v-fyn, and v-fes), except v-abl they are rarely activated in human tumors.

In chronic myeloid leukemia and some lymphoblastic leukemias, the C-abl gene is translocated from its normal abode on chromosome 9 to chromosome 22, here it fuses with part of bcr (break-point cluster region) gene on chromosome 22 and the hybrid gene has potent tyrosine kinase activity.

New evidence suggests that c-abl similar to p53 is activated after DNA damage and hence may play a role in regulating apoptosis.

8.12.4 EGF (Epithelial Growth Factor) (EGF-R, c-erb1-4 o Her-2/neu)

The epithelial growth factor receptor (EGFR) is localized on chromosome 7. It belongs to the erbB family (of tyrosinekinase receptors) comprising the EGFR gene, erbB-1, erbB-3 and erbB-4, and is a transmembrane glucoprotein of 170 kDA. EGFRs or C-erbB-2s play an important role in the transduction of the differentiation, development and emission of the mitogenic signal in normal cells. At different stages of malignant transformation in tissue an anomalous increase of the erbB-1oncogene is produced. According to Werkmeister et al. the gene is 20.2% over expressed in oral carcinomas. Other studies have found an over expression of the EGFR gene in several human cancers, including oral squamous cell carcinoma. The oncogene erbB-2 is localized on the short arm of chromosome 7 and its over expression (14.7% in oral carcinomas) increases metastatic potential. The *ERBB2* gene, also known as *HER-2/neu*, is mapped to region 17q11-q12 (Yamamoto *et al.*, 1986) and is amplified in tumors of various tissues and is a potential marker of prognosis in some them. This gene encodes a transmembrane phosphoglucoprotein (p185) that resembles the epidermal growth factor receptor (EGFR) which acts as a tyrosine kinase receptor, stimulating cell proliferation.

Ultimately all signal transduction pathways enter the nucleus and impact on a large bank of responder genes that orchestrate the cells orderly advance through the mitotic cycle. This process (ie. DNA replication and cell division) is regulated by a family of genes whose products are localized to the nucleus where they control the transcription of growth related genes. The transcription factors contain specific amino acid sequences or motifs that allow them to bind DNA or to dimerize for DNA binding. Examples of such motifs include helix-loop-helix, leucin zipper, zinc finger and homeodomains.

Many of these proteins bind to DNA at specific sites from which they can activate or inhibit transcription of adjacent genes.

A whole host of oncoproteins, including products of the myc, myb, Jun, and fos oncogenes, have been localized to the nucleus of these myc gene is most commonly involved in human tumors.

The c-myc proto-oncogene is expressed in virtually all eukaryotic cells and belongs to the immediate early growth response genes, which are rapidly induced when quiescent cells receive a single to divide.

After translation, c-myc protein is rapidly translocated to the nucleus. Either before of after transport to the nucleus, it forms a heterodiamer with another protein called max. The myc-max heterodimer binds to specific DNA sequences (termed E-boxes) and is a potent transcriptional activator. Mutations that impair the ability of myc to bind to DNA or to max also abolish its oncogenic activity.

mad another member of the myc super family of transcriptional regulators, can also bind max to form a dimer. The max-mad heterodimer functions as a transcription repressor. Thus emerging theme seems to be that the degree of transcriptional activation by c-myc is regulated not only by the levels of myc protein but also by the abundance and availability of max and mad proteins. In this network myc-max favors proliferation, whereas mad-max inhibits cell growth. mad may therefore be considered an antioncogene (tumor suppressor gene).

It is becoming increasingly evident that myc not only controls the cell growth, but also it can drive cell death by apoptosis. Thus, when myc activation occurs in the absence of survival signals (growth factors) cells undergo apoptosis.

Dysregulation of c-myc expression resulting from translocation of gene occurs in Burkett's lymphoma, c-myc is amplified in breast, colon, lung and many other carcinomas. N-myc and L-myc genes are amplified in neuroblastomas and small cell cancers of lungs. The related N-myc and L-myc genes are amplified in neuroblastomas and small cell cancers.

8.12.5 Cyclins (cyclin A, B1, D1, E)

Cyclins are essential in controlling the cell cycle. Their activation triggers the start of the cell cycle and increases replication. A high cdk2 expression is a critical factor in the progression of cancer and can be used as a predictive marker in its prognosis. The cyclin protein D1plays an important role in the later stages of the malignisation process. CD1 has been found to be 39.62% over expressed in oral squamous cell and pharyngeal carcinomas. Cyclin D1 is a 38 kDa protein belonging to the cyclin D family which includes cyclins D2 and D3. The D1 cyclin regulates the transition of cells from the G1 phase to the S phase by phosphorylation of the pRb protein during the G1 phase and the release of the family of E2F transcription factors which in turn lead to the induction of the genes needed for the G1-S cell cycle transition. Cyclin D1 acts by forming the D1 cyclin/cyclin-dependent protein kinase (CDK) complex D1-CDK which activates specific CDKs (CDK4 and CDK6) which allow the cell cycle to progress (Fig below). Amplification of the gene encoding the D1 cyclin is seen in a variety of solid tumors, including breast adenocarcinoma, squamous cell carcinoma of head and neck, esophageal and bladder cancer.

Normal cell proliferation is controlled by growth factors and cytokines that act on the cell membrane by triggering the cascade of biochemical signals (a process called signal transduction). These signals control the genes that regulate cell growth and division. Oncogenes are altered forms of normal cellular genes called proto-oncogenes that are involved in this cascade of events. They may mutate spontaneously through interaction with viruses, chemicals, or by physical means.

These cellular genes were first discovered by the Noble laureate Michael Bishop and Harold Varmus as passengers within the genome of acute transforming retroviruses, which cause rapid induction of tumors in animals and can also transform animal cells in vitro. Molecular dissection of their genomes revealed the presence of unique transforming sequences not found in the genomes of nontransforming retroviruses. Most surprisingly, molecular hybridization revealed that the viral oncogenes were almost identical to sequences found in the normal cellular DNA.

Most surprisingly, molecular hybridization revealed that the v-onc (viral oncogenes) sequences were almost identical to sequences found in the normal cellular DNA. From this evolved the concept that during evolution, retroviral oncogenes were transuduced (captured) by the virus through a chance recombination with the DNA of a (normal) host cell that had been infected by the virus. Because they were discovered initially as viral genes, proto-oncogenes are named after their viral homologs. Each v-oncogene is designated by the oncogene to the virus from which it was isolated. Thus the v-onc contained in feline sarcoma virus is referred to as v-fes, whereas the oncogene in simian sarcoma virus is called v-sis.

V-oncs are not present in several cancer causing RNA viruses. Example is a group of so called slow transforming viruses that cause leukemia's in rodents after a long latent period. The mechanism by which they cause neoplastic transformation implicates proto-oncogenes. Molecular dissection of the cells transformed by these leukemia viruses has revealed that the proviral DNA is always found to be integrated.

(Inserted) near a proto-oncogene. One consequence of proviral insertion near a proto-oncogene is to induce a structural change in the cellular gene, thus converting it in to a cellular oncogene (c-onc). The strong retroviral promoters inserted in the proto-oncogenes lead to dysregulated expression of the cellular gene. This mode of proto-oncogene activation is called insertional mutagenesis.

Oncogenes are altered growth promoting regulatory genes that governs the cells signal transduction pathways and mutations of these genes leads to either overproduction or increased function of the excitatory proteins. Although oncogenes alone are not sufficient to transform epithelial cells, they appear to be important initiators of the process, and are known to cause cellular changes through mutation of only one gene copy.

8.12.6 Bcl2/BAG1

The anti-apoptotic protein Bcl2 is located in the mitochondrial membrane and is regulated by the protein p53. It forms part of the regulatory system that controls the cell cycle and the induction of apoptosis. Apoptosis (from root words meaning "a flling away from") is a relatively distinctive and important mode of cell death that should be differentiated from coagulation necrosis. Apoptosis is responsible for programmed cell death in several important physiologic process including:

The programmed destruction of cells during embryogenesis as occur in implantation, organogenesis and development involution.

Hormone–dependent physiologic involution, such as the endometrium during menstrual cycle or the lacting breast after weaning or pathologic atrophy as in the prostrate after castration.

Cell deletion in proliferating population such as intestinal crypt epithelium or cell death in tumors.

Deleting of auto reactive T cell in the thymus, cell death of cytokine – starved lymphocytes or cell death induced by cytotoxic T cells.

This protein is a key factor in the G1 check point and is therefore the key to the R point. Koontongkaew et al. found this protein to be 58.49% over expressed in the oral carcinomas they studied. Deregulation of the pRb gives rise to aberrations in various cell proteins such as CD1 and CDK4; this mechanism is necessary for the development of oral and pharyngeal cancer.

The signal transducing proteins are oncoproteins that mimic the function of normal cytoplasmic signal transducing proteins.

Such proteins are strategically located on the inner leaflet of the plasma membrane, where they receive signals from outside the cell by activating GFR and transmit them to the cells nucleus. The best and most well studied example of a signal transducing oncoprotein is ras family of guanine triphosphate (GTP) binding proteins.

Most such proteins are strategically located on the inner leaflet of the plasma membrane, where they receive signals from outside the cells (e.g. by activation of growth factor receptors) and transmit them to the cell nucleus. The signal transducing proteins are heterogeneous. The best example of a signal transducing oncoprotein is the ras family of guanine triphosphate (GTP) - binding protein.

The ras proteins were discovered initially in the form of viral oncogenes. Approximately 10 to 20% of all human tumors contain mutated versions of ras proteins.

Mutation of the gene is the single most common abnormality of dominant oncogenes in human tumors. ras plays an important role in mutagenesis induced by growth factors for example, blockade of ras function by microinjection of specific antibodies blocks the proliferative response to EGF, PDGF, and CSF-1.

In the inactive state, ras proteins bind guanosine diphosphate (GDP). When cells are stimulated by growth factors or other receptor ligand interactions, ras becomes activated by exchanging GDP for GTP. The activated ras excites the MAP kinase pathway by recruiting the cytosolic protein raf-1. The MAP kinases activated target nuclear transcription factors and thus promote mitogenesis.

In normal cells, the activated signal transmitting stage of ras protein is transient because its intrinsic GTPase activity hydrolyzes GTP to GDP, thereby returning ras to its quiescent ground state.

The orderly cycling of the ras protein depends on two reactions:

1. Nucleotide exchange (GDP by GTP) which activates ras protein, and
2. GTP hydrolysis, which converts the GTP-bound inactive form.

The removal of GDP and its replacement by GTP during ras activation is catalyzed by a family of guanine nucleotide releasing proteins that are recruited to the cytosolic aspect of activated growth factor receptors by adaptor proteins. The GTPase activity is accelerated by GTPase activating proteins (GAPs). These widely distributed proteins bind to the active ras and augment its GTPase activity by more than 1000-fold leading to rapid hydrolysis of GTP to GDP and termination of signal transduction. Thus, GAPs function as 'brakes' that prevent uncontrolled ras activity.

Mutant ras proteins bind GAP, but their GTPase activity fails to be augmented. Hence the mutant proteins are 'trapped' in their excited GTP-bound form, causing in turn, a pathologic activation of the mitogenic signaling pathway.

ras is also involved in the regulation of cell cycle, as the passage of cells from G0 to the S phase is modulated by a series of proteins called cyclins and cyclin dependent kinases (CDKs). ras controls the levels of CDKs.

To block ras activity, ras must be anchored under the cell membrane close to the cytoplasmic domain of the growth factor receptors. Such anchoring is made possible by attachment of an isoprenyl lipid group to the ras molecule by the enzyme farnesyl transferase. The farnesyl moiety forms the bridge between ras and the lipid membrane. Inhibitors of farnesyl transferase can disable ras by preventing its normal localization.

In addition to ras, several non-receptor associated tyrosine kinases also function in the signal transduction pathways. The mutant forms of non-receptor associated tyrosine kinases are commonly found in the forms of v-oncs in animal retroviruses (e.g. v-abl, v-src, v-fyn, and v-fes), except v-abl they are rarely activated in human tumors.

In chronic myeloid leukemia and some lymphoblastic leukemias, the C-abl gene is translocated from its normal abode on chromosome 9 to chromosome 22, here it fuses with part of bcr (break-point cluster region) gene on chromosome 22 and the hybrid gene has potent tyrosine kinase activity.

New evidence suggests that c-abl similar to p53 is activated after DNA damage and hence may play a role in regulating apoptosis.

8.12.7 EGF (Epithelial Growth Factor) (EGF-R, c-erb1-4 o Her-2/neu)

The epithelial growth factor receptor (EGFR) is localized on chromosome 7. It belongs to the erbB family (of tyrosinekinase receptors) comprising the EGFR gene, erbB-1, erbB-3 and erbB-4, and is a transmembrane glucoprotein of 170 kDA. EGFRs or C-erbB-2s play an important role in the transduction of the differentiation, development and emission of the mitogenic signal in normal cells. At different stages of malignant transformation in tissue an anomalous increase of the erbB-1 oncogene is produced. According to Werkmeister et al. the gene is 20.2% over expressed in oral carcinomas. Other studies have found an over expression of the EGFR gene in several human cancers, including oral squamous cell carcinoma. The oncogene erbB-2 is localized on the short arm of chromosome 7 and its over expression (14.7% in oral carcinomas) increases metastatic potential. The *ERBB2* gene, also known as *HER-2/neu*, is mapped to region 17q11-q12 (Yamamoto *et al.*, 1986) and is amplified in tumors of various tissues and is a potential marker of prognosis in some them. This gene encodes a transmembrane phosphoglucoprotein (p185) that resembles the epidermal growth factor receptor (EGFR) which acts as a tyrosine kinase receptor, stimulating cell proliferation.

Ultimately all signal transduction pathways enter the nucleus and impact on a large bank of responder genes that orchestrate the cells orderly advance through the mitotic cycle. This process (ie. DNA replication and cell division) is regulated by a family of genes whose products are localized to the nucleus where they control the transcription of growth related genes. The transcription factors contain specific amino acid sequences or motifs that allow them to bind DNA or to dimerize for DNA binding. Examples of such motifs include helix-loop-helix, leucin zipper, zinc finger and homeodomains.

Many of these proteins bind to DNA at specific sites from which they can activate or inhibit transcription of adjacent genes.

A whole host of oncoproteins, including products of the myc, myb, Jun, and fos oncogenes, have been localized to the nucleus of these myc gene is most commonly involved in human tumors.

The c-myc proto-oncogene is expressed in virtually all eukaryotic cells and belongs to the immediate early growth response genes, which are rapidly induced when quiescent cells receive a single to divide.

After translation, c-myc protein is rapidly translocated to the nucleus. Either before of after transport to the nucleus, it forms a heterodiamer with another protein called max. The myc-max heterodimer binds to specific DNA sequences (termed E-boxes) and is a potent transcriptional activator. Mutations that impair the ability of myc to bind to DNA or to max also abolish its oncogenic activity.

mad another member of the myc super family of transcriptional regulators, can also bind max to form a dimer. The max-mad heterodimer functions as a transcription repressor. Thus emerging theme seems to be that the degree of transcriptional activation by c-myc is regulated not only by the levels of myc protein but also by the abundance and availability of max and mad proteins. In this network myc-max favors proliferation, whereas mad-max inhibits cell growth. mad may therefore be considered an antioncogene (tumor suppressor gene).

It is becoming increasingly evident that myc not only controls the cell growth, but also it can drive cell death by apoptosis. Thus, when myc activation occurs in the absence of survival signals (growth factors) cells undergo apoptosis.

Dysregulation of c-myc expression resulting from translocation of gene occurs in Burkett's lymphoma, c-myc is amplified in breast, colon, lung and many other carcinomas. N-myc and L-myc genes are amplified in neuroblastomas and small cell cancers of lungs. The related N-myc and L-myc genes are amplified in neuroblastomas and small cell cancers.

This protein is a key factor in the G1 check point and is therefore the key to the R point. Koontongkaew *et al.* found this protein to be 58.49% over expressed in the oral carcinomas they studied. Deregulation of the pRb gives rise to aberrations in various cell proteins such as CD1 and CDK4; this mechanism is necessary for the development of oral and pharyngeal cancer.

The signal transducing proteins are oncoproteins that mimic the function of normal cytoplasmic signal transducing proteins.

Such proteins are strategically located on the inner leaflet of the plasma membrane, where they receive signals from outside the cell by activating GFR and transmit them to the cells nucleus. The best and most well studied example of a signal transducing oncoprotein is ras family of guanine triphosphate (GTP) binding proteins.

Most such proteins are strategically located on the inner leaflet of the plasma membrane, where they receive signals from outside the cells (e.g. by activation of growth factor receptors) and transmit them to the cell nucleus. The signal transducing proteins are heterogeneous. The best example of a signal transducing oncoprotein is the ras family of guanine triphosphate (GTP) - binding protein.

The ras proteins were discovered initially in the form of viral oncogenes. Approximately 10 to 20% of all human tumors contain mutated versions of ras proteins.

Mutation of the gene is the single most common abnormality of dominant oncogenes in human tumors. ras plays an important role in mutagenesis induced by growth factors for example, blockade of ras function by microinjection of specific antibodies blocks the proliferative response to EGF, PDGF, and CSF-1.

In the inactive state, ras proteins bind guanosine diphosphate (GDP). When cells are stimulated by growth factors or other receptor ligand interactions, ras becomes activated by exchanging GDP for GTP. The activated ras excites the MAP kinase pathway by recruiting the cytosolic protein raf-1. The MAP kinases activated target nuclear transcription factors and thus promote mitogenesis.

In normal cells, the activated signal transmitting stage of ras protein is transient because its intrinsic GTPase activity hydrolyzes GTP to GDP, thereby returning ras to its quiescent ground state.

The orderly cycling of the ras protein depends on two reactions:

1. Nucleotide exchange (GDP by GTP) which activates ras protein, and
2. GTP hydrolysis, which converts the GTP-bound inactive form.

The removal of GDP and its replacement by GTP during ras activation is catalyzed by a family of guanine nucleotide releasing proteins that are recruited to the cytosolic aspect of activated growth factor receptors by adaptor proteins. The GTPase activity is accelerated by GTPase activating proteins (GAPs). These widely distributed proteins bind to the active ras and augment its GTPase activity by more than 1000-fold leading to rapid hydrolysis of GTP to GDP and termination of signal transduction. Thus, GAPs function as 'brakes' that prevent uncontrolled ras activity.

Mutant ras proteins bind GAP, but their GTPase activity fails to be augmented. Hence the mutant proteins are 'trapped' in their excited GTP-bound form, causing in turn, a pathologic activation of the mitogenic signaling pathway.

ras is also involved in the regulation of cell cycle, as the passage of cells from G0 to the S phase is modulated by a series of proteins called cyclins and cyclin dependent kinases (CDKs). ras controls the levels of CDKs.

To block ras activity, ras must be anchored under the cell membrane close to the cytoplasmic domain of the growth factor receptors. Such anchoring is made possible by attachment of an isoprenyl lipid group to the ras molecule by the enzyme farnesyl transferase. The farnesyl moiety forms the bridge between ras and the lipid membrane. Inhibitors of farnesyl transferase can disable ras by preventing its normal localization.

In addition to ras, several non-receptor associated tyrosine kinases also function in the signal transduction pathways. The mutant forms of non-receptor associated tyrosine kinases are commonly found in the forms of v-oncs in animal retroviruses (e.g. v-abl, v-src, v-fyn, and v-fes), except v-abl they are rarely activated in human tumors.

In chronic myeloid leukemia and some lymphoblastic leukemias, the C-abl gene is translocated from its normal abode on chromosome 9 to chromosome 22, here it fuses with part of bcr (break-point cluster region) gene on chromosome 22 and the hybrid gene has potent tyrosine kinase activity.

New evidence suggests that c-abl similar to p53 is activated after DNA damage and hence may play a role in regulating apoptosis.

8.12.8 EGF (Epithelial Growth Factor) (EGF-R, c-erb1-4 o Her-2/neu)

The epithelial growth factor receptor (EGFR) is localized on chromosome 7. It belongs to the erbB family (of tyrosinekinase receptors) comprising the EGFR gene, erbB-1, erbB-3 and erbB-4, and is a transmembrane glucoprotein of 170 kDA. EGFRs or C-erbB-2s play an important role in the transduction of the differentiation, development and emission of the mitogenic signal in normal cells. At different stages of malignant transformation in tissue an anomalous increase of the erbB-1 oncogene is produced. According to Werkmeister et al. the gene is 20.2% over expressed in oral carcinomas. Other studies have found an over expression of the EGFR gene in several human cancers, including oral squamous cell carcinoma. The oncogene erbB-2 is localized on the short arm of chromosome 7 and its over expression (14.7% in oral carcinomas) increases metastatic potential. The *ERBB2* gene, also known as *HER-2/neu*, is mapped to region 17q11-q12 (Yamamoto *et al.*, 1986) and is amplified in tumors of various tissues and is a potential marker of prognosis in some them. This gene encodes a transmembrane phosphoglucoprotein (p185) that resembles the epidermal growth factor receptor (EGFR) which acts as a tyrosine kinase receptor, stimulating cell proliferation.

Ultimately all signal transduction pathways enter the nucleus and impact on a large bank of responder genes that orchestrate the cells orderly advance through the mitotic cycle. This process (ie. DNA replication and cell division) is regulated by a family of genes whose products are localized to the nucleus where they control the transcription of growth related genes. The transcription factors contain specific amino acid sequences or motifs that allow them to bind DNA or to dimerize for DNA binding. Examples of such motifs include helix-loop-helix, leucin zipper, zinc finger and homeo-domains.

Many of these proteins bind to DNA at specific sites from which they can activate or inhibit transcription of adjacent genes.

A whole host of oncoproteins, including products of the myc, myb, Jun, and fos oncogenes, have been localized to the nucleus of these myc gene is most commonly involved in human tumors.

The c-myc proto-oncogene is expressed in virtually all eukaryotic cells and belongs to the immediate early growth response genes, which are rapidly induced when quiescent cells receive a single to divide.

After translation, c-myc protein is rapidly translocated to the nucleus. Either before of after transport to the nucleus, it forms a heterodiamer with another protein called max. The myc-max heterodimer binds to specific DNA sequences (termed E-boxes) and is a potent transcriptional activator. Mutations that impair the ability of myc to bind to DNA or to max also abolish its oncogenic activity.

mad another member of the myc super family of transcriptional regulators, can also bind max to form a dimer. The max-mad heterodimer functions as a transcription repressor. Thus emerging theme seems to be that the degree of transcriptional activation by c-myc is regulated not only by the levels of myc protein but also by the abundance and availability of max and mad proteins. In this network myc-max favors proliferation, whereas mad-max inhibits cell growth. mad may therefore be considered an antioncogene (tumor suppressor gene).

It is becoming increasingly evident that myc not only controls the cell growth, but also it can drive cell death by apoptosis. Thus, when myc activation occurs in the absence of survival signals (growth factors) cells undergo apoptosis.

Dysregulation of c-myc expression resulting from translocation of gene occurs in Burkett's lymphoma, c-myc is amplified in breast, colon, lung and many other carcinomas. N-myc and L-myc genes are amplified in neuroblastomas and small cell cancers of lungs. The related N-myc and L-myc genes are amplified in neuroblastomas and small cell cancers.

Apoptosis in oral squamous cell carcinoma is lower in poorly differentiated carcinomas but it is the result of increased antiapototic factors. The Bcl2 family of protein appears to regulate apoptosis via differential homodimension and heterodimension.

High concentrations of Bcl2 may prevent the induction of several forms of apoptosis[18], giving rise to the development of carcinomas, promoting mutations and tumor progression. The function of BAG1 is the opposite to that of Bcl2.

8.12.9 Fas/FasL

These apoptosis mediators belong to the TNF-R family. FasL has been found to be over expressed in oral carcinoma. The absence of Fas receptors indicates poor tumor differentiation.

They have evolved proteins to control the growth of the epithelial cells they infect. This was a necessity since these viruses require a metabolically active, dividing cell in order to complete their life cycle. In particular, the E6 and E7 proteins have the ability to abrogate growth and differentiation controls that would otherwise prevent epithelial cell growth and stymie viral propagation. The "E" designation indicates an early gene, meaning a viral gene that is turned on early in the process of infecting a cell.

The HPV genome typically consists of nine open-reading frame sequences, located on only one of the strands of DNA, and is divided into seven early-phase genes (E) and two late-phase genes (L). The early genes serve to regulate the transcription of DNA, while the late genes encode for proteins involved in viral spread, such as capsid proteins. The E1 and E2 gene products are more specifically involved in regulating the transcription and replication of viral proteins. These different gene regions and gene products provide the basis on which molecular detection methods have been created.

The human papillomavirus genome. The "E" designation indicates an early viral protein which is expressed early in a vegetative infection. Similarly, the "L" designation indicates a late viral gene, usually involved in viral protein coats.

The E7 protein targets the retinoblastoma protein, a critical component of cell cycle control. The retinoblastoma protein (Rb) in the unphosphorylated state binds to and sequesters transcription factors necessary for progression through the cell cycle, particularly E2F and related proteins. This prevents cells from dividing until E2F becomes available in the unbound state, usually by release from Rb. In normal cellular physiology, this release is accomplished by Rb phosphorylation by one of the cyclin-dependent kinases. In the case of a papillomavirus infection, E2F release is due to binding of Rb by viral E7 protein.

E7 Effects on Rb. E7 binding of RB leads to release of sequestered E2F, enabling the cell cycle to progress.

8.12.10 Ki-67/MIB

These two markers, which are monoclonal antibodies, increase when there is tissue proliferation. Ki-67 levels are closely related to the histological degree of carcinoma in oral squamous cells.

The diagnostic accuracy of malignant effusions can be improved by employing various cell proliferation markers. MIB1 monoclonal antibody (Ki67) is present in cycling cells, but not in resting cells. Estimation of the percentage of cells reacting with Ki67 immunocytochemical staining can be performed counting 1000 cells on a consecutive high magnification field.

Saleh et al. have shown a statistically significant correlation between the Ki67 index and cytomorphology of benign (9%), suspicious (19%), and malignant (28%) cells. However, cytomorphology should always remain the basis to differentiate benign from malignant serous effusions and Ki67 stain is a valuable adjunct in difficult cases, acting as a complementary tool to routine cytology.

Cancer is ultimately a disease of genes. In order for cells to start dividing uncontrollably, genes which regulate cell growth must be damaged.

Protooncogenes are genes which promote cell growth and mitosis, a process of cell division, and tumor supressor genes discourage cell growth, or temporarily halts cell division from occuring in order to carry out DNA repair. Typically a series of several mutations to these genes are required before a normal cell transforms in to a cancer cell.

Oral carcinogenesis is a multistep process in which genetic events lead to the disruption of the normal regulatory pathways that control basic cellular functions including cell division, differentiation, and cell death.

Cancer develops through four definable stages: initiation, promotion, progression and malignant conversion. These stages may progress over many years. The first stage, initiation, involves a change in the genetic makeup of a cell. This may occur randomly or when a carcinogen interacts with DNA causing damage. This initial damage rarely results in cancer because the cell has in place many mechanisms to repair damaged DNA. However, if repair does not occur and the damage to DNA is in the location of a gene that regulates cell growth and proliferation, DNA repair, or a function of the immune system, then the cell is more prone to becoming cancerous.

During promotion, the mutated cell is stimulated to grow and divide faster and becomes a population of cells. Eventually a benign tumor becomes evident. In human cancers, hormones, cigarette smoke, or bile acids are substances that are involved in promotion. This stage is usually reversible as evidenced by the fact that lung damage can often be reversed after smoking stops.

The progression phase is less well understood. During progression, there is further growth and expansion of the tumor cells over normal cells. The genetic material of the tumor is more fragile and prone to additional mutations. These mutations occur in genes that regulate growth and cell function such as oncogenes, tumor suppressor genes, and DNA mismatch-repair genes. These changes contribute to tumor growth until conversion occurs, when the growing tumor becomes malignant and possibly metastatic. Many of these genetic changes have been identified in the development of colon cancer and thus it has become a model for studying multi-stage carcinogenesis.

The transformation of normal cells in to malignant cells is dependent on mutations in the genes that control cell cycle progression, leading to the loss of regulatory cell cycle growth signals.

It might then be profitable to list some fundamental principles before we delve in to the details of the molecular basis of cancer.

At the molecular level, progression results from accumulation of genetic lesions that in some instances are favored by defects in D.N.A repair.

Three classes of normal regulatory genes-the growth promoting proto-oncogene, the growth inhibiting cancer suppressor genes (antioncogenes), and genes that regulate programmed cell death, or apoptosis-are the principal targets of genetic damage.

Among the molecular mechanisms involved in the carcinogenesis, defects in the regulation of programmed cell death (apoptosis) may contribute to the pathogenesis and progression of cancer. Dysregulation of oncogenes and tumor suppressor genes involved in apoptosis are also associated with tumor development and progression.

Genes that regulate apoptosis may be dominant, as are proto-oncogene, or they may behave as cancer suppressor genes.

In addition to the three classes of genes mentioned earlier, a fourth category of genes, those that regulate repair of damaged DNA are also pertinent in carcinogenesis.

DNA repair genes affect cell proliferation or survival indirectly by influencing the ability of the organism to repair nonlethal damage in other genes, including proto-oncogenes, tumor suppressor genes, and genes that regulate apoptosis.

The unregulated growth that characterizes cancer is caused by damage to DNA, resulting in mutations to gene that encode for proteins controlling cell division. Many mutation events may be required to transform a normal cell in to a malignant cell. These mutations can be caused by chemicals or physical agents called carcinogens, by close exposure to radioactive materials or by certain viruses that can insert their DNA in to the human genome.

Mutations occur spontaneously, and may be passed down from one generation to the next as a result of mutations within germ line.

The genetic hypothesis of cancer implies that a tumor mass results from the clonal expansion of a single progenitor cell that has incurred the genetic damage. Clonality of tumors is assessed quite readily in women who are heterozygous for polymorphic X-linked markers, such as the enzyme glucose-6-phosphate dehydrogenase (G6PD) or X-linked restriction fragment length polymorphism.

A malignant neoplasm has several phenotypic attributes, such as excessive growth, local invasiveness, and the ability to form distant metastases. These characteristics are acquired in a stepwise fashion, a phenomenon called tumor progression.

Many forms of cancer are associated with exposure to environmental factors such as tobacco smoke, radiation, alcohol and certain viruses.

A variety of agents increase the frequency with which cells are converted to the transformed condition, they are said to be carcinogenic agents. Carcinogens may cause epigenetic changes or may act directly or indirectly to change the genotype of the cells.

Although tobacco is clearly of major aetiological significance (IARC 1984) the failure of overtly malignant lesions to develop in all tobacco users and the development of oral cancer in all tobacco users and the development of oral cancer in persons with no history of tobacco use suggests that the genesis of oral cancer may also involve other unidentified environmental and host factors.

8.13 Tumour Growth Markers

8.13.1 Nuclear cell proliferation antigens

These are nuclear proteins associated with DNA-polymerase. They appear in the final phase of G1 and in the S phase. In addition, they are thought to form part of the D-cdk cyclin complex, where they are involved in phases of the cell cycle. They are indicative of cell proliferation[51]. *P120* This is a new component of the catenin family. It is a protein associated with nuclear proliferation in the early stages of the S phase, and is localized next to the centromere of the long arm of chromosome 11. Alterations of the E-cadherin-p120 complex may play an important role in tumor progression. A loss of expression of this complex indicates that the neoplasia is in a state of progression.

Oral leukoplakia with dysplasia is a well-recognized precursor of invasive SCC of the oral cavity. The percentage of leukoplakic lesions that progress to invasive SCC is directly related to the severity of the dysplasia, ranging from under 5% for leukoplakias with mild to moderate dysplasia to as high as 43% for leukoplakias containing severe dysplasia/carcinoma in situ.

This lesion is defined as a white patch or plaque that cannot be diagnosed as any other disease and is not associated with any mechanical or chemical irritant except for the use of tobacco. Histologically, the lesions are sub classified according to the degree of dysplasia and growth rate changes. Because premalignant lesions of different anatomical sites are characterized by increased cell proliferation that usually parallels the degree of dysplasia, cell proliferation markers may be used to assess the type and degree of oral premalignant lesions. Thus, immunohistochemically detectable proliferation markers could be of great value in predetecting lesion behavior and may serve as surrogate end point biomarkers in cancer chemoprevention to evaluate possible regression or improvement in abnormal features in the tissues of subjects at increased risk. The assessment of surrogate end point biomarkers is possible because invasive epithelial neoplasms are known to be preceded by intraepithelial proliferations with a spectrum of cellular abnormalities extending from mild dysplasia to carcinoma in situ.

A number of polypeptide growth factors that stimulate proliferation of normal cells and many are suspected to play a role in tumoriogenesis. Mutations of genes that encode growth factors can render them oncogenic. Such is the case with the proto-oncogenes c-sis, which encode the ß chain of platelet derived growth factor (PDGF).

This oncogene was first discovered in the guise of the viral oncogene contained in v-sis. Autocrine loop is considered to be an important element in the pathogenesis of tumor.

More commonly, products of other oncogenes such as ras (that lie alone the signal transduction pathway) cause over expression of growth factor genes, thus forcing the cells to secrete large amounts of growth factors, such as transforming growth factor (TGFἀ). This growth factor is related to epidermal growth factor (EGF) and induces proliferation by binding to the EGF receptor TGFά is often detected in carcinomas that express high levels of EGF receptors.

Despite extensive documentation of growth factor-medicated autocrine stimulation of transformed cells, increased growth factor production is not sufficient for neoplastic transformation. Extensive cell proliferation, contributes, to the malignant phenotype by increasing the risk of spontaneous or induced mutations in the cell population.

Growth factor receptors are next group in the sequence of signal transduction.

The oncogenic versions of these receptors are associated with persistent dimerization and activation without binding to the growth factor.

Hence the mutant receptors deliver continuous mitogenic signals to the cell.

Growth factor receptors are activated in human tumors by several mechanisms.

1. Mutations
2. Gene rearrangements
3. Over expression

The ret proto-oncogene, a receptor tyrosine kinase, exemplifies oncogenic conversion via mutations and gene rearrangements.

Point mutations in the ret proto-oncogene are associated with dominantly inherited MEN types 2A and 2B. In MEN 2A, point mutations in the extra cellular domain cause constitutive dimerization and activation, whereas in MEN 2B, point mutations in the cytoplasmic catalytic domain activate the receptor. In all these familial tumors, the affected individuals inherit the ret mutations in the germ line.

In myeloid leukemias, the gene encoding the colony stimulating factor1 (CSF1) receptor has been detected. In certain chronic myelomonocytic leukemias with the t (12:9) translocation, the entire cytoplasmic domain of the PDGF receptor is fused with a segment of the ETS family transcription factor, resulting in permanent dimerization of the PDGF receptor.

Three members of the EGF receptor family are the ones most commonly involved. The normal form of c-erb B1, the EGF receptor gene, is over expressed in squamous cell carcinomas of lung, and less commonly in carcinomas of urinary bladder, gastrointestinal tract and astrocytomas. This increased receptor expression results from gene amplification. The c-erb B2 gene (also called c-neu), the second member of the EGF receptor family, is amplified in a high percentage of human Aden carcinomas arising within the breast, ovary, lung, stomach and salivary glands. A third member of the EGF receptor family, c-erb B3 is also over expressed in breast cancers.

8.13.2 AgNOR (argyrophilic nucleolar organiser region) associated proteins

The AgNOR proteins have been defined as loops of nuclear DNA that code for ribosomal DNA. They are argyrophilic and serve as an indicator of nuclear proliferation. Although the quantification and distribution of AgNOR are subjective and non diagnostic parameters of specific lesions, they are useful as a complement to histopathological study in terms of identifying the degree of any cell and nuclear alterations. It is the only marker of this group to show an important association with prognosis and may be indicative of the degree of malignancy.

The signal transducing proteins are oncoproteins that mimic the function of normal cytoplasmic signal transducing proteins.

Such proteins are strategically located on the inner leaflet of the plasma membrane, where they receive signals from outside the cell by activating GFR and transmit them to the cells nucleus. The best and most well studied example of a signal transducing oncoprotein is ras family of guanine triphosphate (GTP) binding proteins.

Most such proteins are strategically located on the inner leaflet of the plasma membrane, where they receive signals from outside the cells (e.g. by activation of growth factor receptors) and transmit them to the cell nucleus. The signal transducing proteins are heterogeneous. The best example of a signal transducing oncoprotein is the ras family of guanine triphosphate (GTP) - binding protein.

The ras proteins were discovered initially in the form of viral oncogenes. Approximately 10 to 20% of all human tumors contain mutated versions of ras proteins.

Mutation of the gene is the single most common abnormality of dominant oncogenes in human tumors. ras plays an important role in mutagenesis induced by growth factors for example, blockade of ras function by microinjection of specific antibodies blocks the proliferative response to EGF, PDGF, and CSF-1.

In the inactive state, ras proteins bind guanosine diphosphate (GDP). When cells are stimulated by growth factors or other receptor ligand interactions, ras becomes activated by exchanging GDP for GTP. The activated ras excites the MAP kinase pathway by recruiting the cytosolic protein raf-1. The MAP kinases activated target nuclear transcription factors and thus promote mitogenesis.

In normal cells, the activated signal transmitting stage of ras protein is transient because its intrinsic GTPase activity hydrolyzes GTP to GDP, thereby returning ras to its quiescent ground state.

The orderly cycling of the ras protein depends on two reactions:

1. Nucleotide exchange (GDP by GTP) which activates ras protein, and
2. GTP hydrolysis, which converts the GTP-bound inactive form.

The removal of GDP and its replacement by GTP during ras activation is catalyzed by a family of guanine nucleotide releasing proteins that are recruited to the cytosolic aspect of activated growth factor receptors by adaptor proteins. The GTPase activity is accelerated by GTPase activating proteins (GAPs). These widely distributed proteins bind to the active ras and augment its GTPase activity by more than 1000-fold leading to rapid hydrolysis of GTP to GDP and termination of signal transduction. Thus, GAPs function as 'brakes' that prevent uncontrolled ras activity.

Mutant ras proteins bind GAP, but their GTPase activity fails to be augmented. Hence the mutant proteins are 'trapped' in their excited GTP- bound form, causing in turn, a pathologic activation of the mitogenic signaling pathway.

ras is also involved in the regulation of cell cycle, as the passage of cells from G0 to the S phase is modulated by a series of proteins called cyclins and cyclin dependent kinases (CDKs). ras controls the levels of CDKs.

To block ras activity, ras must be anchored under the cell membrane close to the cytoplasmic domain of the growth factor receptors. Such anchoring is made possible by attachment of an isoprenyl lipid group to the ras molecule by the enzyme farnesyl transferase. The farnesyl moiety forms the bridge between ras and the lipid membrane. Inhibitors of farnesyl transferase can disable ras by preventing its normal localization.

In addition to ras, several non-receptor associated tyrosine kinases also function in the signal transduction pathways. The mutant forms of non-receptor associated tyrosine kinases are commonly found in the forms of v-oncs in animal retroviruses (e.g. v-abl, v-src, v-fyn, and v-fes), except v-abl they are rarely activated in human tumors.

In chronic myeloid leukemia and some lymphoblastic leukemias, the C-abl gene is translocated from its normal abode on chromosome 9 to chromosome 22, here it fuses with part of bcr (break-point cluster region) gene on chromosome 22 and the hybrid gene has potent tyrosine kinase activity.

New evidence suggests that c-abl similar to p53 is activated after DNA damage and hence may play a role in regulating apoptosis.

Ultimately all signal transduction pathways enter the nucleus and impact on a large bank of responder genes that orchestrate the cells orderly advance through the mitotic cycle. This process (ie. DNA replication and cell division)is regulated by a family of genes whose products are localized to the nucleus where they control the transcription of growth related genes. The transcription factors contain specific amino acid sequences or motifs that allow them to bind DNA or to dimerize for DNA binding. Examples of such motifs include helix-loop-helix, leucin zipper, zinc finger and homeodomains.

Many of these proteins bind to DNA at specific sites from which they can activate or inhibit transcription of adjacent genes.

A whole host of oncoproteins, including products of the myc, myb, Jun, and fos oncogenes, have been localized to the nucleus of these myc gene is most commonly involved in human tumors.

The c-myc proto-oncogene is expressed in virtually all eukaryotic cells and belongs to the immediate early growth response genes, which are rapidly induced when quiescent cells receive a single to divide.

After translation, c-myc protein is rapidly translocated to the nucleus. Either before of after transport to the nucleus, it forms a heterodiamer with another protein called max. The myc-max heterodimer binds to specific DNA sequences (termed E-boxes) and is a potent transcriptional activator. Mutations that impair the ability of myc to bind to DNA or to max also abolish its oncogenic activity.

mad another member of the myc super family of transcriptional regulators, can also bind max to form a dimer. The max-mad heterodimer functions as a transcription repressor. Thus emerging theme seems to be that the degree of transcriptional activation by c-myc is regulated not only by the levels of myc protein but also by the abundance and availability of max and mad proteins. In this network myc-max favors proliferation, whereas mad-max inhibits cell growth. mad may therefore be considered an antioncogene (tumor suppressor gene).

It is becoming increasingly evident that myc not only controls the cell growth, but also it can drive cell death by apoptosis. Thus, when myc activation occurs in the absence of survival signals (growth factors) cells undergo apoptosis.

Dysregulation of c-myc expression resulting from translocation of gene occurs in Burkett's lymphoma, c-myc is amplified in breast, colon, lung and many other carcinomas. N-myc and L-myc genes are amplified in neuroblastomas and small cell cancers of lungs. The related N-myc and L-myc genes are amplified in neuroblastomas and small cell cancers.

8.13.3 Skp2 (S-phase kinase-interacting protein 2)

Sphase Kinase interacting protein2 (Skp2), an Fbox protein, is required for the ubiquitination and consequent degradation of p27. It is well known that reduce dexpression of p27 is frequently observed invarious cancers including oral squamous cell carcinoma an disduetoan enhancement of its protein degradation.

Skp2 over expression was frequently observed in squamous cell carcinoma.

SCFSkp2 was identified as the E3 ubiquitin ligase that targets p27 for ubiquitination. SCF complexes represent an evolutionarily conserved class of E3 enzymes containing four subunits: Skp1, Cul1, one of many F box proteins, and Roc1/Rbx1. Skp2, an F box protein, is required for the ubiquitination and consequent degradation of p27 both in vivo and in vitro. In addition, in vitro ubiquitination of recombinant p27 can be induced by the addition of purified Skp2 and cyclin E/cyclin-dependent kinase (Cdk) 2 or cyclin A/Cdk2 complexes to G1 cell extracts. Antisense oligonucleotides to Skp2 or over expression of a dominant-negative Skp2 mutant stabilize p27protein in vivo. These findings indicate that Skp2 is specifically required for p27 ubiquitination and that Skp2 is a rate-limiting component of the machinery that ubiquitinates and degrades phosphorylated p27. Skp2 is frequently over expressed in tumor cell lines, and forced expression of Skp2 in quiescent fibroblasts induces DNAsynthesis.

Skp2 over expression was frequently observed in oral squamous cell carcinoma. Furthermore, Skp2 expression increases significantly during malignant progression from epithelial dysplasia to invasive oral squamous cell carcinoma.

Importantly, Skp2 over expression was well correlated with down-regulation of p27 protein in oral squamous cell carcinoma. It is well known that reduced expression of p27 is frequently found in various cancers, and the lack of p27 is suggested to be due to an enhancement of its degradation. p27 is Cdk inhibitor and mediates G1arrest induced by transforming growth factor-h, contact inhibition, or serum deprivation in epithelial cell lines. The increase in the cellular abundance of p27 upon induction of cell quiescence is primarily due to a decrease in the rate of its degradation. p27 is polyubiquitinated both in vivo and in vitro and a lower amount of p27 ubiquitinating activity is present in proliferating cells compared with quiescent cells. In fact, aggressive human cancers such as colon cancers, lymphomas, and astrocytic brain tumors express low levels of p27 because of its decreased stability. It is also found that reduced expression of p27 was shown in 87% of oral squamous cell carcinoma cases and was well correlated with its malignancy including metastasis and poor prognosis. These previous findings suggest that enhanced p27 degradation observed in oral squamous cell carcinoma might be due to increased levels of Skp2, and that Skp2 over expression may play an important role for the development of oral squamous cell carcinoma.

High expression of this protein is linked to a decrease in p27 and has been associated with poor prognosis.

8.13.4 HSP27 and 70 (heat shock proteins)

These appear to be associated with mutations of the p53 gene. The HSP27 protein is found in normal mucosa and small tumors. High levels of HSP70 have been detected in oral squamous cell carcinomas. Both proteins interact with Bcl2, lending support to the proliferation effect.

The functions of such genes is to arrest the progression of cell cycle in order to carry out DNA repair, preventing mutations from passed on to daughter cells. The tumor suppressor genes are most often inactivated by point mutations, deletions and rearrangements in both gene copies. However, a mutation can damage the tumor suppressor gene itself. The invariable consequence of this is that DNA repair is hindered or inhibited. DNA damage accumulates without repair, inevitably leading to cancer.

There has been much research on the tumor suppressor gene p53. The p53 protein blocks cell division at the G1 to S boundary, stimulates DNA repair after DNA damage, and also induces apoptosis. These functions are achieved by the ability of p53 to modulate the expression of several genes. The p53 protein transcriptionally activates the production of the p21 protein, encoded by the WAF1/CIP gene, p21 being an inhibitor of cyclin and cyclin dependant kinase complexes. p21 transcription is activated by wild-type p53 but not mutant p53. However, WAF1/CIP expression is also induced by p53 independent pathways such as growth factors, including platelet derived growth factor (PDGF), fibroblast growth factor (FGF), and transforming growth factor ß (TGF-ß). Wild-type p53 has a very short half life (four to five minutes), whereas mutant forms of protein are more stable, with a six hour half life.

Mutation of p53 occurs either as a point mutation, which results in a structurally altered protein that sequesters the wild-type protein, thereby inactivating its suppressor activity, or by deletion, which leads to a reduction or loss of p53 expression and protein function.

The tumor suppressor gene p53 is known to be mutated in approximately 70% of adult solid tumors.

p53 has been shown to be functionally inactivated in oral tumors, and restoration of p53 in oral cancer lines and tumors induced in animal models has been shown to reverse the malignant phenotype. Smoking and tobacco use have been associated with the mutation of p53 in head and neck cancers.

Other tumor suppressor genes include doc-1, the retinoblastoma gene, and APC. The doc-1 gene is mutated in malignant oral keratinocytes, leading to a reduction of expression and protein function. The precise function of the Doc-1 protein in oral carcinogenisis is unclear, but it is very similar to a gene product induced in mouse fibroblasts by tumor necrosis factor α(TNF-α). Normally, TNF-α decreases proliferation and increases differentiation, and has been shown in oral squamous cell carcinoma cell lines to be responsible, either alone or in combination with interferons α or γ, for antiproliferative activity.

8.13.5 Telomerase

This is a DNA protein structure located at the end of eukaryote chromosomes. Telomeric activity is essential for controlling the unlimited potential for division and the immortality of eukaryote cells. This activity, which is not detected in normal somatic cells, can be evaluated in biopsied tissue. As in other tumors, this activity is used as a marker in the diagnosis of pre-neoplastic or neoplastic oral mucosa lesions, as 80–90% of such tumors have high levels of telomeric expression, particularly of the hTERT sub-unit (catalytic activity). Detection (in particular, of the hTERT sub-unit) may be useful as an additional diagnostic marker.

In cellular aging normal cells become arrested in a terminally nondividing state, known as cellular senescence. But with each cell division there is some shortening of specialized structure (telomerase) at the ends of chromosomes. Once the telomerase are shortened beyond a certain point, the loss of telomerase function leads to end-to-end chromosome fusion and cell death. Thus telomerase shortening is believed to be a clock that counts cell divisions. In germ cells, telomerase shortening is prevented by the sustained function of enzymes telomerase, thus explaining the ability of these cells to self replicate extensively. This enzyme is absent from most somatic cells and hence they suffer progressive loss of telomerase.

Telomerase activity has been detected in the vast majority of human tumors and in those that lack telomerase, other telomerase–lengthening mechanisms have been found.

Allelic loss of 3p and 9p and other regions containing tumor suppressor genes has also been reported in precursor lesions of oral cancer showing varying degrees of dysplasia compared with normal epithelium. Allelic loss or imbalance at p53, DCC (deleted in colon carcinoma), and regions at 3p21.30–22, and 3p12.1–13 were reported, with LOH at DCC shown to occur in areas of dysplasia adjacent to infiltrating carcinoma. This suggested that loss at this locus might be a later event, whereas LOH at 3p and p53 were more frequent in those dysplastic Chromosome breakpoints are frequently seen in centromeric regions of chromosomes 1, 3, 8, 14, 15, 1p22, 11q13, and 19p13. Because genes bcl-1, int-2, and hst-1 have been mapped to 11q13 and n-ras to 11q13, it has been suggested that activation of these oncogenes is the result of these cytogenic alterations.

Approximately two thirds of all head and neck cancers contain a deleted region in chromosome 9p21–22. The cyclin dependent kinases inhibitor 2/multiple tumor suppressor gene 1 (CDKN2/MTSI) has been mapped to this chromosome region, and inactivation of its protein product p16^{INK4} by mutation and deletion has been found in 10% and 33% of head and neck squamous carcinomas, respectively, along with frequent inactivation of p16 in oral premalignant lesions. This suggests an important role for this gene in the early stages of oral carcinogenesis. Cyclins, cyclin dependant kinases (CDKs), and cyclin dependent kinases inhibitors regulate progress through key transitions in the cell cycle. p16^{INK4} binds to and inhibits phosphorylation of pRb by the cyclin dependent kinases CDK4 and CDK6.

Other proteins that regulate crucial checkpoints in the cell cycle, and which are important contributors to increased cell proliferation, include cyclin D, E, and A, which regulate the G1 to S phase transition, and cyclin B, which regulates the G2 to M transition.

The cyclin D1 gene is frequently over expressed in oral cancers as a result of amplification of the 11q13 region. Overexpressin of cyclin a has been reported in oral carcinomas, with the increase in expression being associated with tumor grade. Cyclin B was also reported to be over expressed, with increased cytoplasmic staining compared with nuclear staining in normal cells. Cyclin B1 binds to protein kinases p34^{cdc2} in the cytoplasm of the dividing cells, and the complex is transported to the nucleus at the G2 to M transition. This suggests that frequent abnormalities in cyclin B/p34^{cdc2} kinetics in oral carcinomas lead to deregulation of the G2 to M transition.

Oral carcinoma is the most common malignancy of orofacial region though strictly speaking carcinoma means malignancies of epithelial tissue origin.

The malignant neoplasms of epithelial cell origin, derived from any of the three germ layers, are called carcinomas. Thus cancer arising in the epidermis of ectodermal origin is a carcinoma, as is a cancer arising in the mesodermally derived cells of the renal tubules and the endodermally derived cells of the lining of gestrointestinal tract.

Carcinomas may be further qualified. One with a glandular growth pattern microscopically is termed an adenocarcinoma, and one producing recognizable squamous cells arising in any epithelium of the body is termed a squamous cell carcinoma.

Carcinogenesis is a multistep process at both phenotypic and genetic level means the creation of cancer. It includes the process of derangement of the rate of cell division due to damage to D.N.A. so cancer is ultimately a disease of genes. Cancer is caused by a series of mutations. Each mutation alters the behavior of the cell.

The transformation of normal cells in to malignant cells is dependent on mutations in the genes that control cell cycle progression, leading to the loss of regulatory cell cycle growth signals.

It might then be profitable to list some fundamental principles before we delve in to the details of the molecular basis of cancer.

At the molecular level, progression results from accumulation of genetic lesions that in some instances are favored by defects in D.N.A repair.

Three classes of normal regulatory genes-the growth promoting proto-oncogene, the growth inhibiting cancer suppressor genes (antioncogenes), and genes that regulate programmed cell death, or apoptosis-are the principal targets of genetic damage.

Among the molecular mechanisms involved in the carcinogenesis, defects in the regulation of programmed cell death (apoptosis) may contribute to the pathogenesis and progression of cancer. Dysregulation of oncogenes and tumor suppressor genes involved in apoptosis are also associated with tumor development and progression.

Genes that regulate apoptosis may be dominant, as are proto-oncogene, or they may behave as cancer suppressor genes.

In addition to the three classes of genes mentioned earlier, a fourth category of genes, those that regulate repair of damaged DNA are also pertinent in carcinogenesis.

DNA repair genes affect cell proliferation or survival indirectly by influencing the ability of the organism to repair nonlethal damage in other genes, including proto-oncogenes, tumor suppressor genes, and genes that regulate apoptosis.

The unregulated growth that characterizes cancer is caused by damage to DNA, resulting in mutations to gene that encode for proteins controlling cell division. Many mutation events may be required to transform a normal cell in to a malignant cell. These mutations can be caused by chemicals or physical agents called carcinogens, by close exposure to radioactive materials or by certain viruses that can insert their DNA in to the human genome.

Mutations occur spontaneously, and may be passed down from one generation to the next as a result of mutations within germ line.

The genetic hypothesis of cancer implies that a tumor mass results from the clonal expansion of a single progenitor cell that has incurred the genetic damage. Clonality of tumors is assessed quite readily in women who are heterozygous for polymorphic X-linked markers, such as the enzyme glucose-6-phosphate dehydrogenase (G6PD) or X-linked restriction fragment length polymorphism.

A malignant neoplasm has several phenotypic attributes, such as excessive growth, local invasiveness, and the ability to form distant metastases. These characteristics are acquired in a stepwise fashion, a phenomenon called tumor progression.

Many forms of cancer are associated with exposure to environmental factors such as tobacco smoke, radiation, alcohol and certain viruses.

A variety of agents increase the frequency with which cells are converted to the transformed condition, they are said to be carcinogenic agents. Carcinogens may cause epigenetic changes or may act directly or indirectly to change the genotype of the cells.

Although tobacco is clearly of major aetiological significance (IARC 1984) the failure of overtly malignant lesions to develop in all tobacco users and the development of oral cancer in all tobacco users and the development of oral cancer in persons with no history of tobacco use suggests that the genesis of oral cancer may also involve other unidentified environmental and host factors.

Several studies have identified specific genetic alterations in oral carcinomas and in premalignant lesions of the oral cavity. Recently, using comparative genomic hybridization on primary oral carcinomas, Bockmuhl and colleagues reported deletions of chromosome 3p, 5q, and 9p with 3q gain in well differentiated tumors, whereas in poorly differentiated tumors deletions of 4q, 8p, 11q, 13q, 18q, and 21q and gains in 1p, 11q, 13, 19, and 22q were identified, thus suggesting an association with tumor progression. With the development of molecular techniques, such as micro satellite assays and restriction fragment length polymorphism, it has been shown that allelic imbalance of chromosomal 9p is the most common chromosomal arm loss in oral carcinoma.

Several oncogenes have also been implicated in oral carcinogenesis. Aberrant expression of the proto-oncogene epidermal growth factor receptor (EGFR/c-erb 1), members of the ras gene family, c-myc, int-2, hst-1, PRAD-1, and bcl-1 is believed to contribute towards cancer development.

8.14 Tumour Suppression and Carcinogenesis

8.14.1 Bax

This is a p53 co-factor which acts in the induction of apoptosis; it is induced by p53. Low levels of Bax have been linked to poor prognosis in squamous cell carcinoma.

Bax–Bax homodimers promote cell death. Few studies have been undertaken to investigate apoptosis in oral cancers, although Jordan and colleagues demonstrated that Bcl-2 was present in poorly differentiated cancers, whereas Bax was present in differentiated oral cancers. In a more recent study in oral cancers from an Asian population, low concentrations of Bax were demonstrated, with a high concentration of Bcl-2, irrespective of tumour differentiation. In this regard, in oral primary and metastatic squamous carcinomas an increased expression of $Bclx_s$, Bik, and Bax—proteins that stimulate apoptosis—in contrast to those that inhibit apoptosis, including Bcl_2, Mcl-1, and Bcl-x.

8.14.2 Dendritic cells (DC)

Dendritic cells (DCs) are immune cells and form part of the mammalian immune system. Their main function is to process antigen material and present it on the surface to other cells of the immune system, thus functioning as antigen-presenting cells.

Dendritic cells are present in small quantities in tissues that are in contact with the external environment, mainly the skin (where they are often called Langerhans cells) and the inner lining of the nose, lungs, stomach and intestines. They can also be found at an immature state in the blood. Once activated, they migrate to the lymphoid tissues where they interact with T cells and B cells to initiate and shape the adaptive immune response. At certain development stages they grow branched projections, the dendrites, that give the cell its name. However, these do not have any special relation with neurons, which also possess similar appendages. Immature dendritic cells are also called veiled cells, in which case they possess large cytoplasmic 'veils' rather than dendrites.

There are two types of cells with dendritic morphology that are functionally quite different. Both have numerous fine dendritic cytoplasmic processes, from which they derive their name. One type is called interdigating dendritic cells or just dendritic cells. These cells are nonphagocytic and they express high levels of MHC class II molecules as well as the costimulary molecules B7-1 and B7- 2. Dendritic cells are widely distributed. They are found in lymphoid tissue and in the interstitium of many nonlymphoid organs. The other type of cells with dendritic morphology are present in the germinal centres of lymphoid follicles in the spleen and lymph nodes and are hence called follicular dendritic cells.

These are capable of generating an important anti-tumor response. Their over expression is indicative of a good prognosis.

Dendritic cells (DCs) are the most potent antigen-presenting cells (APCs). They play a key role in antitumor immunity and actively participate in the generation of tumor specific, antitumor effector Tcells. DCs are particularly efficient in executing the major histo-compatibility complex (MHC) class I dependent pathway of antigen presentation, using either self-derived peptides or peptides derived through exogenous pathways. In tumor-bearing hosts, DCs uptake, process, and cross-present tumor-associated antigens (TAAs) or memory T cells. The MHC class I-restricted presentation by DCs of internalized and processed peptides leads to the generation of tumor specific effector T cells capable of recognizing and eliminating tumor cells. The presence on DCs of co stimulatory molecules (CD40, CD80, and CD86) and of MHC class I and II molecules also is crucial for the generation of cytolytic T lymphocytes (CTL).

The number of S-100 positive DCs present in the tumor is by far the strongest independent predictor of overall survival as well as DFS and TTR in patients with oral carcinoma compared with well established prognostic factors as disease stage or lymph node involvement. In contrast, when another immuno histochemical marker, p55, identify tumor-infiltrating DCs, a significant but substantially weaker correlation with survival. The p55, which marks a subpopulation of DCs expressing actin bundling protein, is not as useful in predicting survival as S-100, may be a reflection of the functional heterogeneity of DCs within the tumor. It implies that one subset of DCs (S-100 positive) has a different biologic importance than another subset of DCs (p55positive). In the tumor, DCs are likely to internalize TAAs for subsequent presentation to T cells in the regional lymph nodes. Thus, only a subset of DCs present in the tumor may be recognized by the p55 antibody, and, functionally, this subset appears to be less critical for the generation of antitumor responses than the S-100 positive DC subset.

8.14.3 Zeta chains

These have recently been identified as part of the T-cell receptor, which is involved in tumor defense. Lack of zeta chain expression in tumors has been associated with reduced survival.

Alterations in the expression of signaling molecules in T cells of patients with cancer have been reported by many investigators. These alterations, including a decrease in expression or absence of the signal-transducing Ã‡chain, are thought to be responsible for functional impairments of immune cells at the tumor site and in the peripheral circulation of patients with malignancies. A significant decrease or absence of the f chain in T cells incubated with human tumor cells *in vitro suggest that this alteration may be tumor induced.*

The mechanism(s) responsible for a decrease in expression are unknown, and various explanations have been advanced, including the possibility that it is an artifact induced by monocytes during isolation of peripheral blood mononuclear cells. A decreased expression of the f chain has been linked to the process of apoptosis induced in T cells by tumor-derived or other apoptotic signals. Overall, however, the biological. The signal transducing proteins are oncoproteins that mimic the function of normal cytoplasmic signal transducing proteins.

Such proteins are strategically located on the inner leaflet of the plasma membrane, where they receive signals from outside the cell by activating GFR and transmit them to the cells nucleus. The best and most well studied example of a signal transducing oncoprotein is ras family of guanine triphosphate (GTP) binding proteins.

Most such proteins are strategically located on the inner leaflet of the plasma membrane, where they receive signals from outside the cells (e.g. by activation of growth factor receptors) and transmit them to the cell nucleus. The signal transducing proteins are heterogeneous. The best example of a signal transducing oncoprotein is the ras family of guanine triphosphate (GTP) - binding protein.

The ras proteins were discovered initially in the form of viral oncogenes. Approximately 10 to 20% of all human tumors contain mutated versions of ras proteins.

Mutation of the gene is the single most common abnormality of dominant oncogenes in human tumors. ras plays an important role in mutagenesis induced by growth factors for example, blockade of ras function by microinjection of specific antibodies blocks the proliferative response to EGF, PDGF, and CSF-1.

In the inactive state, ras proteins bind guanosine diphosphate (GDP). When cells are stimulated by growth factors or other receptor ligand interactions, ras becomes activated by exchanging GDP for GTP. The activated ras excites the MAP kinase pathway by recruiting the cytosolic protein raf-1. The MAP kinases activated target nuclear transcription factors and thus promote mitogenesis.

In normal cells, the activated signal transmitting stage of ras protein is transient because its intrinsic GTPase activity hydrolyzes GTP to GDP, thereby returning ras to its quiescent ground state.

The orderly cycling of the ras protein depends on two reactions:

1. Nucleotide exchange (GDP by GTP) which activates ras protein, and
2. GTP hydrolysis, which converts the GTP-bound inactive form.

The removal of GDP and its replacement by GTP during ras activation is catalyzed by a family of guanine nucleotide releasing proteins that are recruited to the cytosolic aspect of activated growth factor receptors by adaptor proteins. The GTPase activity is accelerated by GTPase activating proteins (GAPs). These widely distributed proteins bind to the active ras and augment its GTPase activity by more than 1000-fold leading to rapid hydrolysis of GTP to GDP and termination of signal transduction. Thus, GAPs function as 'brakes' that prevent uncontrolled ras activity.

Mutant ras proteins bind GAP, but their GTPase activity fails to be augmented. Hence the mutant proteins are 'trapped' in their excited GTP-bound form, causing in turn, a pathologic activation of the mitogenic signaling pathway.

ras is also involved in the regulation of cell cycle, as the passage of cells from G0 to the S phase is modulated by a series of proteins called cyclins and cyclin dependent kinases (CDKs). ras controls the levels of CDKs.

To block ras activity, ras must be anchored under the cell membrane close to the cytoplasmic domain of the growth factor receptors. Such anchoring is made possible by attachment of an isoprenyl lipid group to the ras molecule by the enzyme farnesyl transferase. The farnesyl moiety forms the bridge between ras and the lipid membrane. Inhibitors of farnesyl transferase can disable ras by preventing its normal localization.

In addition to ras, several non-receptor associated tyrosine kinases also function in the signal transduction pathways. The mutant forms of non-receptor associated tyrosine kinases are commonly found in the forms of v-oncs in animal retroviruses (e.g. v-abl, v-src, v-fyn, and v-fes), except v-abl they are rarely activated in human tumors.

In chronic myeloid leukemia and some lymphoblastic leukemias, the C-abl gene is translocated from its normal abode on chromosome 9 to chromosome 22, here it fuses with part of bcr (break-point cluster region) gene on chromosome 22 and the hybrid gene has potent tyrosine kinase activity.

New evidence suggests that c-abl similar to p53 is activated after DNA damage and hence may play a role in regulating apoptosis.

The patients with stage III or IV tumors and absent or low expression of the f chain in TIL have extremely poor survival compared to patients with either normal f chain expression or early-stage oral SCC.

8.15 Markers of Tumour Invasion and Metastatic Potential

8.15.1 MMPs (matrix-metallo proteases)

Oral squamous cell carcinomas are highly invasive lesions that destroy adjacent tissues and invade bone and muscle, which is most likely the result of matrix metalloproteinase (MMP) activity. MMPs are zinc-dependent endopeptidases, which collectively are capable of degrading virtually all ECM components. This family of matrix-degrading enzymes participates in tissue remodeling processes under both physiological and pathological conditions including morphogenesis, angiogenesis, wound healing, arthritis, and tumor invasion. They are subdivided based on substrate specificity and structural characteristics into the collagenases (MMP-1, MMP-8, and MMP-13), gelatinases (MMP-2 and MMP-9), stromelysins (MMP-3, MMP-10, and MMP-11), Matrilysin (MMP-7), and membrane-anchored metalloproteinases (MT1-MMP, MT2-MMP, MT3-MMP, MT4-MMP, and MT5-MMP).

Over expression of several of these MMPs, either in tumor cells or in tumor-associated fibroblasts, has been linked with increased invasive and metastatic behavior in many different types of cancers including oral SCC.

The expression of the zinc metallo enzymes has been found in oral squamous cell carcinoma and is associated with the tumor stage.

Tumor invasive process is thought to involve the multiple proteolytic enzyme matrix metalloproteinases. Among the MMPs, MMP-2 and MMP-9 have been thought to be key enzymes in this process, because they degrade type IV collagen, which is one of the important components of extra cellular matrix. Membrane type 1-MMP (MT1-MMP) has been originally identified as an activator of Pro-MMP-2. On the other hand, tissue inhibitor of metalloproteinase 2 (TIMP-2) is an inhibitor of MMP-2.

Matrix metalloproteinase (MMP)-2 and MMP-9 are considered to play an important role in the metastasis of malignant tumors. Membrane type 1-MMP (MT1-MMP) and tissue inhibitor of metalloproteinase 2 (TIMP-2) are essential factors for the activation of pro-MMP-2. There are some reports about expressions of MMP family in relationship to clinical features of head and neck squamous cell carcinoma (SCC).

O-charoenrat *et al.* reported that the mRNA level of MMP-9 correlated with advanced tumor stage and lymph node status at diagnosis in head and neck SCC. Riedel *et al.* demonstrated that MMP-9 expression did not correlate with tumor and lymph node stages, but correlated with worse survival in head and neck SCC. Kurahara et al. reported that MMP-9 expression correlated with tumor invasion and lymph node involvement in oral SCC. Therefore, MMP-9 expression may be a useful marker for predicting tumor metastasis and for prognosis in head and neck SCC including early-stage oral SCC.

In contrast to MMP-9 expression, clinical association and prognostic values of MMP-2 and MT1-MMP are still controversial in head and neck SCC. In head and neck SCC recruited from different anatomical sites, O-charoenrat *et al.* reported that mRNA level of MMP-2 correlated with lymph node involvement. Imanishi et al. reported that expression of MT1-MMP correlated with lymph node metastasis but that of MMP-2 did not. In oral SCC, Kusukawa et al. reported that expression of pro MMP-2 correlated with lymph node involvement. Kurahara et al. reported that expressions of MMP-2 and MT1-MMP correlated with tumor invasion and lymph node involvement. However, these previous studies analyzed only the status at diagnosis and did not examine the predictive values for tumor recurrence and prognosis. Recently, Yoshizaki *et al.* reported that marked expressions of MMP-2 and

MT1-MMP correlated with lymph node recurrence and worse survival as well as with clinical stage at the time of diagnosis in tongue SCC. Therefore, the predictive value of MMP-2 and MT1-MMP expressions for tumor recurrence or metastases and prognosis may vary according to primary sites, tumor sizes, and/or lymph node status at the time of diagnosis in head and neck SCC. In general expression of MMP-9 and marked expression of TIMP-2 are associated with regional lymph node and/or distant metastasis and poor prognosis.

8.15.2 Integrins

A family of transmembrane, cell surface receptors composed of two subunits: alpha and beta. Expression of the integrin $\alpha v\beta 6$ is induced during tumor genesis and epithelial repair. Various studies have shown that the integrin $\alpha v\beta 6$ is expressed in squamous cell carcinoma of the oral cavity.

Hamidi et al.[47] found that 41% of leukoplakias expressed the integrin $\alpha v\beta 6$, which may be associated with processes of epithelial repair, inflammation or malignant transformation. The expression of this integrin seems to be necessary, but not sufficient, to produce this transformation.

8.15.3 Cadherins and catenins

They are a family of glycoprotein that act as glue between epithelial cells. Loss of cadherins can favor the malignant phenotype by allowing easy disaggregation of cells, which can then invade locally or metastasize. Recent studies indicate that like many other tumor suppressor genes germ line mutation of E-cadherins gene can predispose to familial gastric carcinoma. Their main function is maintaining polarity and tissue architecture. The expression of these molecules is inversely proportional to tumor differentiation.

The molecular basis of reduced E-cadherin expression is varied. In a small proportion of cases, there are mutations in the E-caderin gene (located on 16q): in others cancer, E-caderin expression is reduced secondary to mutation in the catenins genes. Catenins bind to the intracellular portion of cadherins and stabilize their expression.

This protein is a key factor in the G1 check point and is therefore the key to the R point. Koontongkaew et al. found this protein to be 58.49% over expressed in the oral carcinomas they studied. Deregulation of the pRb gives rise to aberrations in various cell proteins such asCD1 and CDK4; this mechanism is necessary for the development of oral and pharyngeal cancer.

The signal transducing proteins are oncoproteins that mimic the function of normal cytoplasmic signal transducing proteins.

Such proteins are strategically located on the inner leaflet of the plasma membrane, where they receive signals from outside the cell by activating GFR and transmit them to the cells nucleus. The best and most well studied example of a signal transducing oncoprotein is ras family of guanine triphosphate (GTP) binding proteins.

Most such proteins are strategically located on the inner leaflet of the plasma membrane, where they receive signals from outside the cells (e.g. by activation of growth factor receptors) and transmit them to the cell nucleus. The signal transducing proteins are heterogeneous. The best example of a signal transducing oncoprotein is the ras family of guanine triphosphate (GTP) - binding protein.

The ras proteins were discovered initially in the form of viral oncogenes. Approximately 10 to 20% of all human tumors contain mutated versions of ras proteins.

Mutation of the gene is the single most common abnormality of dominant oncogenes in human tumors. ras plays an important role in mutagenesis induced by growth factors for example, blockade of ras function by microinjection of specific antibodies blocks the proliferative response to EGF, PDGF, and CSF-1.

In the inactive state, ras proteins bind guanosine diphosphate (GDP). When cells are stimulated by growth factors or other receptor ligand interactions, ras becomes activated by exchanging GDP for GTP. The activated ras excites the MAP kinase pathway by recruiting the cytosolic protein raf-1. The MAP kinases activated target nuclear transcription factors and thus promote mitogenesis.

In normal cells, the activated signal transmitting stage of ras protein is transient because its intrinsic GTPase activity hydrolyzes GTP to GDP, thereby returning ras to its quiescent ground state.

The orderly cycling of the ras protein depends on two reactions:

1. Nucleotide exchange (GDP by GTP) which activates ras protein, and
2. GTP hydrolysis, which converts the GTP-bound inactive form.

The removal of GDP and its replacement by GTP during ras activation is catalyzed by a family of guanine nucleotide releasing proteins that are recruited to the cytosolic aspect of activated growth factor receptors by adaptor proteins. The GTPase activity is accelerated by GTPase activating proteins (GAPs). These widely distributed proteins bind to the active ras and augment its GTPase activity by more than 1000-fold leading to rapid hydrolysis of GTP to GDP and termination of signal transduction. Thus, GAPs function as 'brakes' that prevent uncontrolled ras activity.

Mutant ras proteins bind GAP, but their GTPase activity fails to be augmented. Hence the mutant proteins are 'trapped' in their excited GTP-bound form, causing in turn, a pathologic activation of the mitogenic signaling pathway.

ras is also involved in the regulation of cell cycle, as the passage of cells from G0 to the S phase is modulated by a series of proteins called cyclins and cyclin dependent kinases (CDKs). ras controls the levels of CDKs.

To block ras activity, ras must be anchored under the cell membrane close to the cytoplasmic domain of the growth factor receptors. Such anchoring is made possible by attachment of an isoprenyl lipid group to the ras molecule by the enzyme farnesyl transferase. The farnesyl moiety forms the bridge between ras and the lipid membrane. Inhibitors of farnesyl transferase can disable ras by preventing its normal localization.

In addition to ras, several non-receptor associated tyrosine kinases also function in the signal transduction pathways. The mutant forms of non-receptor associated tyrosine kinases are commonly found in the forms of v-oncs in animal retroviruses (e.g. v-abl, v-src, v-fyn, and v-fes), except v-abl they are rarely activated in human tumors.

In chronic myeloid leukemia and some lymphoblastic leukemias, the C-abl gene is translocated from its normal abode on chromosome 9 to chromosome 22, here it fuses with part of bcr (break-point cluster region) gene on chromosome 22 and the hybrid gene has potent tyrosine kinase activity.

New evidence suggests that c-abl similar to p53 is activated after DNA damage and hence may play a role in regulating apoptosis.

8.15.4 Desmoplakin/placoglobin

Low expression of these molecules has been associated with distant metastasis.

Ultimately all signal transduction pathways enter the nucleus and impact on a large bank of responder genes that orchestrate the cells orderly advance through the mitotic cycle. This process (ie. DNA replication and cell division) is regulated by a family of genes whose products are localized to the nucleus where they control the transcription of growth related genes. The transcription factors contain specific amino acid sequences or motifs that allow them to bind DNA or to dimerize for DNA binding. Examples of such motifs include helix-loop-helix, leucin zipper, zinc finger and homeodomains.

Many of these proteins bind to DNA at specific sites from which they can activate or inhibit transcription of adjacent genes.

A whole host of oncoproteins, including products of the myc, myb, Jun, and fos oncogenes, have been localized to the nucleus of these myc gene is most commonly involved in human tumors.

The c-myc proto-oncogene is expressed in virtually all eukaryotic cells and belongs to the immediate early growth response genes, which are rapidly induced when quiescent cells receive a single to divide.

After translation, c-myc protein is rapidly translocated to the nucleus. Either before of after transport to the nucleus, it forms a heterodiamer with another protein called max. The myc-max heterodimer binds to specific DNA sequences (termed E-boxes) and is a potent transcriptional activator. Mutations that impair the ability of myc to bind to DNA or to max also abolish its oncogenic activity.

mad another member of the myc super family of transcriptional regulators, can also bind max to form a dimer. The max-mad heterodimer functions as a transcription repressor. Thus emerging theme seems to be that the degree of transcriptional activation by c-myc is regulated not only by the levels of myc protein but also by the abundance and availability of max and mad proteins. In this network myc-max favors proliferation, whereas mad-max inhibits cell growth. mad may therefore be considered an antioncogene (tumor suppressor gene).

It is becoming increasingly evident that myc not only controls the cell growth, but also it can drive cell death by apoptosis. Thus, when myc activation occurs in the absence of survival signals (growth factors) cells undergo apoptosis.

Dysregulation of c-myc expression resulting from translocation of gene occurs in Burkett's lymphoma, c-myc is amplified in breast, colon, lung and many other carcinomas. N-myc and L-myc genes are amplified in neuroblastomas and small cell cancers of lungs. The related N-myc and L-myc genes are amplified in neuroblastomas and small cell cancers.

8.15.5 Ets-1

A protooncogene that acts as a transcription factor. It has been linked to tumor stage and lymphatic metastases.

Identification of molecular changes characteristic of development and progression of oral cancer are of paramount importance for effective intervention. Stromelysin 3 (MMP11) is a unique matrix metalloproteinase shown to have dual function during cancer progression. The transcription factor Ets-1 is important proangiogenic factors in oral cancer.

Head and neck tumorigenesis is a multistep process. Squamous cell carcinomas (SCC) of the oral cavity, a subgroup of head and cancer, are clinically preceded by precancerous lesions, often leukoplakia, with histological evidence of hyperplasia or dysplasia. Only a small proportion (5–10%) of the early dysplastic lesions progress to SCC over a period of 10 years.

The transcription factor Ets-1 is induced in endothelial cells by angiogenic factors, VEGF, and basic fibroblast growth factor. Ets-1 promotes neovascularization by inducing angiogenesis-related genes such as *MMPs* and *Integrin β3*. Ets-1 expression has been reported in a number of human tumors, including oral carcinomas, astrocytomas, gastric carcinomas, and mammary carcinomas. Strikingly, Ets-1 antisense oligonucleotides.Abrogate the invading phenotype and VEGF-induced migration of endothelial cells *in vitro*. The gene for the endothelial cell-specific Flt1/VEGFR1 also contains an Ets-1-responsive element in its promoter, and mutational disruption of this Ets-1 binding site results in a decrease of VEGFR1 transcription. Ets factors have also been shown to interact and cooperate with other transcription factors in the activation of promoters of genes encoding extra cellular matrix proteases. Ets-1 over expression promotes endothelial cell invasiveness and expression of matrix proteases as well as integrin ß3; conferring an angiogenic phenotype to these cells.

Table 8.1 Association of Micro Vessel Density with the Expression of Ets-1, VEGF, and ST3 in Oral Precancerous and Cancerous Lesions

	Total Cases	iMVD, mean $\pm$ SD, (mm^2/HPF)	P
Precancerous lesions	90	22.8 ± 21.6	
Ets-1$^+$	56	26.14 ± 21.54	0.05
Ets-1$^-$	34	17.26 ± 21.54	
VEGF$^+$	59	27.32 ± 23.5	0.001
VEGF$^-$	31	14.16 ± 14.26	
ST3$^+$	48	27.35 ± 26.24	0.026
ST3$^-$	42	17.6 ± 13.17	
SCCs	220	25.4 ± 18.3	
Ets-1$^+$	170	27.02 ± 19.05	0.016
Ets-1$^-$	50	19.9 ± 14.2	
VEGF$^+$	167	26.85 ± 19.36	0.014
VEGF$^-$	53	20.9 ± 13.5	
ST3$^+$	148	27.21 ± 18.64	0.036
ST3$^-$	72	21.72 ± 17.07	

Abbreviations: iMVD, intraoral micro vessel density

Table 8.2 Association of Alterations in ST3, VEGF, and Ets-1 Expression with Transition from Normal to Precancerous Stage

Logistic Regression Analysis					95.0% CI	
Variable	Wald	*df*	*P*	OR	Lower	Upper
Univariate						
VEGF+	18.326	1	0.000	4.026	2.128	7.618
Ets-1$^+$	10.578	1	0.001	2.800	1.505	5.207
ST3$^+$	14.042	1	0.000	3.486	1.814	6.698
ST3$^+$/VEGF$^+$	16.684	1	0.000	5.806	2.497	13.504
Ets-1$^+$/VEGF$^+$	15.786	1	0.000	3.936	2.002	7.737
Ets-1$^+$/ST3$^+$	16.981	1	0.000	5.188	2.371	11.353
Multivariate						
VEGF$^+$	10.069	1	0.002	2.970	1.516	5.819
Ets-1$^+$/ST3$^+$	9.786	1	0.001	3.701	1.630	8.403

Abbreviation: CI, confidence interval.

Table 8.3 Association of Alterations in ST3, VEGF, and Ets-1 Expression with Transition from Precancerous to Cancerous Stage

Logistic regression analysis					95.0% CI	
Variable	Wald	*df*	*P*	OR	Lower	Upper
Univariate						
Ets-1$^+$	7.181	1	0.007	2.064	1.215	3.507
VEGF$^+$	3.432	1	0.064	1.656	.971	2.822
ST3$^+$	5.278	1	0.022	1.799	1.090	2.968
Ets-1$^+$/VEGF$^+$	6.614	1	0.010	1.926	1.169	3.174
Ets-1$^+$/ST3$^+$	8.019	1	0.005	2.053	1.248	3.376
ST3$^+$/VEGF$^+$	8.083	1	0.004	2.068	1.253	3.412
Ets-1$^+$/ST3$^+$/VEGF$^+$	9.230	1	0.002	2.202	1.323	3.663
Multivariate						
ST3$^+$/VEGF$^+$	8.083	1	0.004	2.068	1.253	3.412

Abbreviation: CI, confidence interval.

8.16 Cell Surface Markers

8.16.1 Carbohydrates and Antigen

Increased levels of the mucin complex at the cell surface are associated with a heightened degree of dysplasia. Cell-surface carbohydrates with blood group antigen activity are widely distributed in human tissues. The term 'histo-blood group antigens' has been suggested for blood group antigens located on cells other than erythrocytes e.g *CD57 antigen* which is found in the membrane of lymphoid and neural cells. The percentage of CD57 lymphocytes is increased in oral leukoplakias with moderate or severe dysplasia compared with normal tissue. Histo-blood group antigens of the ABH, Lewis, and T/Tn systems are seen at the surfaces of epithelial cells in oral squamous epithelium. During cellular differentiation in stratified squamous epithelium, there is a sequential elongation of the terminal carbohydrate chain of precursors of histo-blood group antigens by the action of gene-encoded glycosyl transferases.

During malignant development, the synthesis of histo-blood group antigens is disturbed possibly due to aberrant expression of the glycosyl transferases. Almost 30 years ago, Dabelsteen and Pindborg (1973) showed that histo-blood group antigen A was lost in oral carcinomas. Further, in oral epithelial dysplasias, there was a loss of the normally expressed histo-blood group antigens (A or B) in the spinous cell layer, and an increased number of epithelial cell layers stained for the precursor molecule (H-antigen), which is normally expressed only in the parabasal cells. In normal epithelium, histo-blood group antigen Le^y is present on parabasal cells, whereas in epithelial dysplasias the expression of Le^y is seen in cell surfaces of the superficial spinous cells, possibly reflecting a lack of normal epithelial differentiation. A similar pattern of expression of simple mucin type carbohydrate antigens (T/Tn) has been reported in oral leukoplakias and erythroplakias.

The *Histocompatibility antigen (HLA)* is a molecules which form the class-I immunohisto compatibility complex play a highly important role in immunity. The class-II HLA antigen is expressed in some oral carcinomas, and more commonly in those with little differentiation.

Some of the aberrant expression patterns referred to above were seen in pre-malignant lesions without epithelial dysplasia. suggesting that histo-blood group antigen changes appear early in the development of malignancy. However, only in a very limited number of cases have the histo-blood group antigen changes been related directly to the ultimate fate of the lesions (cancer/non-cancer). It shows pre-malignant lesions that later developed into cancer exhibited a loss of histo-blood group antigen A years before malignant transformation.

8.17 Intracellular Markers

8.17.1 Cytokeratins

Cytokeratins are intermediate filament keratins found in the intra cytoplasmic cytoskeleton of epithelial tissue. There are two types of cytokeratins: the low weight, acidic type I cytokeratins and the high weight, basic or neutral type II cytokeratins. Cytokeratins are usually found in pairs comprising a type I cytokeratin and a type II cytokeratin.

The subsets of cytokeratins which an epithelial cell expresses depends mainly on the type of epithelium, the moment in the course of terminal differentiation and the stage of development. Thus this specific cytokeratin fingerprint allows the classification of all epithelia upon their cytokeratin expression profile. Furthermore this applies also to the malignant counterparts of the epithelia (carcinomas), as the cytokeratin profile tends to remain constant when an epithelium undergoes malignant transformation.

The cytokeratins are encoded by a family encompassing 30 genes. Among them, 20 are epithelial genes and the resting 10 are specific for trichocytes.

All cytokeratin chains are composed of a central α-helix-rich domain (with a 50–90% sequence identity among cytokeratins of the same type and around 30% between cytokeratins of different type) with non-α-helical N- and C-terminal domains. The α-helical domain has 310–150 amino acids and comprises four segments in which a seven-residue pattern repeats. Into this repeated pattern, the first and fourth residues are hydrophobic and the charged residues show alternate positive and negative polarity, resulting in the polar residues being located on one side of the helix. This central domain of the chain provides the molecular alignment in the keratin structure and makes the chains form coiled dimers in solution.

The end-domain sequences of type I and II cytokeratin chains contain in both sides of the rod domain the subdomains V1 and V2, which have variable size and sequence. The type II also presents the conserved subdomains H1 and H2, encompassing 36 and 20 residues respectively. The subdomains V1 and V2 contain residues enriched by glycines and/or serines, the former providing the cytokeratin chain a strong insoluble character and facilitating the interaction with other molecules. These terminal domains are also important in the defining the function of the cytokeratin chain characteristic of a particular epithelial cell type.

Two dimers of cytokeratin groups into a keratin tetramer by anti-parallel binding. This cytokeratin tetramer is considered to be the main building block of the cytokeratin chain. By head-to-tail linking of the cytokeratin tetramers, the protofilaments are originated, which in turn intertwine in pairs to form protofibrils. Four protofibrils give place to one cytokeratin filament.

In the cytoplasm, the keratin filaments can form a complex network which extends from the surface of the nucleus to the cell membrane. Numerous accessory proteins are involved in the genesis and maintenance of such structure.

This association between the plasma membrane and the nuclear surface provides important implications for the organization of the cytoplasm and cellular communication mechanisms. Apart from the relatively static functions provided in terms of supporting the nucleus and providing tensile strength to the cell, the cytokeratin networks undergo rapid phosphate exchanges mediated depolymerization, with important implications in the more dynamic cellular processes such as mitosis and post-mitotic period, cell movement and differentiation.

Cytokeratins interact with desmosomes and hemidesmosomes, thus collaborating to cell-cell adhesion and basal cell-underlying connective tissue connection.

The intermediate filaments of the eukaryotic cytoskeleton, which the cytokeratins are one of its three components, have been probed to associate also with the spectrin complex protein network that underlies the cell membrane.

There are 19 cytokeratins, divided into two sub-families. Changes in the expression of these proteins cannot be considered predictive of the development of dysplasia. The malignisation of oral lesions is associated with the disappearance of cytokeratins. Research has shown that the expression of CK19 in the suprabasal cell layer of the oral mucosa can be used as a diagnostic marker of pre-cancerous oral lesions; CK19 expression has also been localized in the early stages of carcinogenesis.

HNSCCs represent about 4% of all human cancers in the world, and the incidence of this type of malignancy is expected to increase in the future. The morbidity and mortality from HNSCC are significant. and current therapeutic approaches have not improved survival. Five-year survival rates in patients with HNSCC can be enhanced by early diagnosis, but this has been limited by a lack of sufficiently sensitive and objective histopathological markers for detecting premalignant changes.

Cytokeratins, a group of approximately 20 different proteins that are assembled into a network of intermediate filaments, exhibit distinct patterns of expression in specific epithelial tissues. The patterns of keratin expression in normal epithelia and the change in their expression in premalignant lesions and carcinomas have suggested possibilities for improving diagnosis. Changes in the expression of certain cytokeratins have also been detected in premalignant and malignant head and neck epithelial lesions.

The mucosal lining of most of the human oral epithelium is nonkeratinized. However, this epithelium can undergo abnormal keratinization in premalignant (e.g., leukoplakia) and malignant lesions, and this is accompanied by changes in the expression of cytokeratins and other biomarkers. For example, CK 1, a marker of suprabasal keratinocytes in keratinized squamous epithelium, has-been detected in abnormally keratinizing oral epithelial cells and CKI9, which is expressed in simple epithelium and the basal cells of squamous epithelium, was found to be expressed in premalignant oral lesions. Likewise, CK8, which is expressed in simple epithelium, is not expressed in normal oral mucosa cells but may be expressed in carcinomas. In contrast, CKI3 is expressed in normal nonkeratinized oral epithelium as well as in head and neck cancers.

Intermediary filaments, like cytokeratins are essential intracellular components, underlying or reflecting distinct cellular properties and differentiation stages in epithelial organs. The proteins of the cytokeratin family are epithelium specific expressed as low and high-molecular weight, respectively acid and basic polypeptides. Squamous, stratified epithelium is usually characterized by the expression of CK 5, which is found mainly in the basal cell layers and Ck 5 is associated with the proliferative potential of these cells. The intermediary cell layers show an additional expression of Ck's 1 and 10, which are regarded as signs of cellular differentiation. In contrast, glandular epithelia reveal the expression of low molecular weight cytokeratins Ck's 8/18 and 19 as typical features–expression pattern of Ck 8/18 is rather uncommon in mature squamous epithelium. Ck 19 expressed heterogeneously in the basal cell layers of stratified squamous epithelium. Suprabasal expression of Ck 19 seems to be correlated with premalignant transformation in oral epithelium.

The expression of high molecular weight cytokeratins, especially Ck 5 is a hallmark of squamous epithelium and is predominantly seen in the basal layers of stratified epithelium. This cell layer is regarded as the anatomical localization of tissue specific stem/progenitor cells. Stem cells of stratified epithelium have been described as the major cellular targets for cancer causing mutations and therefore might give in a long term rise to the development of SCC's.

Co expression of cytokeratin has been observed in tumor cells of a variety of salivary gland neoplasms such as AdCC, pleomorphic adenomas, and carcinomas in pleomorphic adenoma. This co expression has been found in human normal salivary glands, but only in a limited area where myoepithelial cells and basal cells of the epithelium are present. Immunohistochemical studies showed nonhomogeneous expression of intermediate filaments in AdCC, where staining for cytokeratin was more striking in luminal cells (type B) than peripheral cells.

8.18 Markers of Anomalous Keratinisation

8.18.1 Filagrins

These proteins, rich in histamine, are found in the granular and corneal layers of the normal epithelium. They are responsible for aggregating keratin between the filaments in the final stages of keratinocyte differentiation. In oral leukoplakias, filagrins appear in the corneal layer, while in oral carcinomas they form keratin pearls. Their expression is thought to be independent of the degree of atypical histology.

The signal transducing proteins are oncoproteins that mimic the function of normal cytoplasmic signal transducing proteins.

Such proteins are strategically located on the inner leaflet of the plasma membrane, where they receive signals from outside the cell by activating GFR and transmit them to the cells nucleus. The best and most well studied example of a signal transducing oncoprotein is ras family of guanine triphosphate (GTP) binding proteins.

Most such proteins are strategically located on the inner leaflet of the plasma membrane, where they receive signals from outside the cells (e.g. by activation of growth factor receptors) and transmit them to the cell nucleus. The signal transducing proteins are heterogeneous. The best example of a signal transducing oncoprotein is the ras family of guanine triphosphate (GTP) - binding protein.

The ras proteins were discovered initially in the form of viral oncogenes. Approximately 10 to 20% of all human tumors contain mutated versions of ras proteins.

Mutation of the gene is the single most common abnormality of dominant oncogenes in human tumors. ras plays an important role in mutagenesis induced by growth factors for example, blockade of ras function by microinjection of specific antibodies blocks the proliferative response to EGF, PDGF, and CSF-1.

In the inactive state, ras proteins bind guanosine diphosphate (GDP). When cells are stimulated by growth factors or other receptor ligand interactions, ras becomes activated by exchanging GDP for GTP. The activated ras excites the MAP kinase pathway by recruiting the cytosolic protein raf-1. The MAP kinases activated target nuclear transcription factors and thus promote mitogenesis.

In normal cells, the activated signal transmitting stage of ras protein is transient because its intrinsic GTPase activity hydrolyzes GTP to GDP, thereby returning ras to its quiescent ground state.

The orderly cycling of the ras protein depends on two reactions:

1. Nucleotide exchange (GDP by GTP) which activates ras protein, and
2. GTP hydrolysis, which converts the GTP-bound inactive form.

The removal of GDP and its replacement by GTP during ras activation is catalyzed by a family of guanine nucleotide releasing proteins that are recruited to the cytosolic aspect of activated growth factor receptors by adaptor proteins. The GTPase activity is accelerated by GTPase activating proteins (GAPs). These widely distributed proteins bind to the active ras and augment its GTPase activity by more than 1000-fold leading to rapid hydrolysis of GTP to GDP and termination of signal transduction. Thus, GAPs function as 'brakes' that prevent uncontrolled ras activity.

Mutant ras proteins bind GAP, but their GTPase activity fails to be augmented. Hence the mutant proteins are 'trapped' in their excited GTP-bound form, causing in turn, a pathologic activation of the mitogenic signaling pathway.

ras is also involved in the regulation of cell cycle, as the passage of cells from G0 to the S phase is modulated by a series of proteins called cyclins and cyclin dependent kinases (CDKs). ras controls the levels of CDKs.

To block ras activity, ras must be anchored under the cell membrane close to the cytoplasmic domain of the growth factor receptors. Such anchoring is made possible by attachment of an isoprenyl lipid group to the ras molecule by the enzyme farnesyl transferase. The farnesyl moiety forms the bridge between ras and the lipid membrane. Inhibitors of farnesyl transferase can disable ras by preventing its normal localization.

In addition to ras, several non-receptor associated tyrosine kinases also function in the signal transduction pathways. The mutant forms of non-receptor associated tyrosine kinases are commonly found in the forms of v-oncs in animal retroviruses (e.g. v-abl, v-src, v-fyn, and v-fes), except v-abl they are rarely activated in human tumors.

In chronic myeloid leukemia and some lymphoblastic leukemias, the C-abl gene is translocated from its normal abode on chromosome 9 to chromosome 22, here it fuses with part of bcr (break-point cluster region) gene on chromosome 22 and the hybrid gene has potent tyrosine kinase activity.

New evidence suggests that c-abl similar to p53 is activated after DNA damage and hence may play a role in regulating apoptosis.

8.18.2 Intercellular substance antigen

This is partially or totally absent in 92% of oral leukoplakias with dysplasia and in 26% of leukoplakias without dysplasia. The loss of expression of this antigen is observed in 95% of oral carcinomas.

While proto-oncogenes encode proteins that promote cell growth, the products of tumor suppressor genes apply breaks to cell proliferation. In normal cells, these genes modulate growth promoting signals, transcription, DNA repair, and replication.

Like proto-oncogenes, their products act at crucial points in the cell cycle to maintain homeostasis.

The physiologic function of these genes is to regulate cell growth not to prevent tumor formation. Because the loss of these genes is a key event in many human tumors. Approximately 60% of retinoblastoma's are sporadic and the remaining 40% are familial, with the predisposition to develop the tumor being transmitted as an autosomal dominant trait. To explain the familial and sporadic occurrence of an apparently identical tumor, Knudson proposed "Two hit" hypothesis of oncogenesis. He suggested that in hereditary cases, one genetic change (first hit) is inherited from an affected parent and is therefore present in all somatic cells of the body, whereas the second mutation "second hit" occurs in one of the many retinal cells. Knudson's hypothesis can now be formulated in more precise terms:

- The mutations required to produce retinoblastoma involve the Rb gene, located on chromosome 13q14. In some cases, the genetic damage is large enough to be visible in the form of a deletion of 13q14.
- Both normal alleles of the Rb locus must be inactivated (two hits) for the development of retinoblastoma. In familial cases, children are born with one normal and one defective copy of the Rb gene. They lose the intact copy in the retinoblasts through some form of somatic mutation (point mutation, interstitial deletion of 13q14, or even complete loss of the normal chromosome 13). In sporadic cases, both normal Rb alleles are lost by somatic mutation in one of the retinoblasts. A retinal cell that has lost both normal copies of the Rb gene gives rise to cancer.
- Patients with familial retinoblastoma are also at greatly increased risk of developing osteosarcoma and other soft tissue sarcomas. Inactivation of the Rb locus noted in adenocarcinoma of the breast, small cell carcinoma of lung, and bladder carcinoma. Because the Rb gene is associated with

cancer when both normal copies are lost, it is sometimes referred to as a recessive cancer gene.

Tumor suppressor genes code for antiproliferation signals and proteins that suppress mitosis and cell growth. Generally tumor suppressors are transcription factors that are activated by cellular stress or DNA damage. Often DNA damage will cause the presence of free floating genetic material as well as other signs, and will trigger enzymes and pathways which lead to the activation of tumor suppressor genes.

The functions of such genes is to arrest the progression of cell cycle in order to carry out DNA repair, preventing mutations from passed on to daughter cells. The tumor suppressor genes are most often inactivated by point mutations, deletions and rearrangements in both gene copies. However, a mutation can damage the tumor suppressor gene itself. The invariable consequence of this is that DNA repair is hindered or inhibited. DNA damage accumulates without repair, inevitably leading to cancer.

There has been much research on the tumor suppressor gene p53. The p53 protein blocks cell division at the G1 to S boundary, stimulates DNA repair after DNA damage, and also induces apoptosis. These functions are achieved by the ability of p53 to modulate the expression of several genes. The p53 protein transcriptionally activates the production of the p21 protein, encoded by the WAF1/CIP gene, p21 being an inhibitor of cyclin and cyclin dependant kinase complexes. p21 transcription is activated by wild-type p53 but not mutant p53. However, WAF1/CIP expression is also induced by p53 independent pathways such as growth factors, including platelet derived growth factor (PDGF), fibroblast growth factor (FGF), and transforming growth factor ß(TGF-ß). Wild-type p53 has a very short half life (four to five minutes), whereas mutant forms of protein are more stable, with a six hour half life.

Mutation of p53 occurs either as a point mutation, which results in a structurally altered protein that sequesters the wild-type protein, thereby inactivating its suppressor activity, or by deletion, which leads to a reduction or loss of p53 expression and protein function.

The tumor suppressor gene p53 is known to be mutated in approximately 70% of adult solid tumors.

p53 has been shown to be functionally inactivated in oral tumors, and restoration of p53 in oral cancer lines and tumors induced in animal models has been shown to reverse the malignant phenotype. Smoking and tobacco use have been associated with the mutation of p53 in head and neck cancers.

Other tumor suppressor genes include doc-1, the retinoblastoma gene, and APC. The doc-1 gene is mutated in malignant oral keratinocytes, leading to a reduction of expression and protein function. The precise function of the Doc-1 protein in oral carcinogenisis is unclear, but it is very similar to a gene product induced in mouse fibroblasts by tumor necrosis factor α(TNF-α). Normally, TNF-α decreases proliferation and increases differentiation, and has been shown in oral squamous cell carcinoma cell lines to be responsible, either alone or in combination with interferons α or γ, for antiproliferative activity.

8.18.3 Nuclear analysis

Sudbo et al. argue that DNA analysis constitutes an important advance in the evaluation of the risk of oral cancer in patients with leukoplakias. Therefore, DNA is a powerful predictor of the risk of a lesion's malignant transformation. One of the most sensitive methods for studying clonal changes in tumors and premalignant lesions is analysis based on the polymerase chain reaction (PCR). The advantage of this procedure is that it requires a small amount of DNA. The analysis can be performed with cells scraped from suspicious surfaces and enables much information to be obtained through what is a non-invasive technique.

The parameters evaluated in nuclear analysis include:

1. DNA ploidy state (of chromosomal pairing), which reflects the risk of oral cancer:

   (a) Anaploidy: high risk
   (b) Tetraploidy: intermediate risk
   (c) Diploidy: low risk

   As a guideline, 32% of oral leukoplakias and 45% of squamous cell carcinomas have anaploid nuclei. Anaploid nuclei are found in 29% of leukoplakias without dysplasia, in 22% of leukoplakias with mild dysplasia, and in 67% of leukoplakias with severe dysplasia. Therefore, it can be said that molecular information enables the evaluation of the risk of oral cancer to be redefined and serves as a treatment guide in the case of lesions such as leukoplakia. In other words, anaploid oral leukoplakias require more aggressive treatments in order to prevent them becoming more malignant.

2. Chromosomal polysomy: This determines genetic instability. Kim et al. reported extensive chromosomal polysomyin areas classified as high risk

of malignisation compared with low-risk areas. These polysomies are much more numerous in dysplasic epithelia compared with hyperplasic epithelial cells.

8.19 Arachidonic Acid Products

Levels of lipoxygenase metabolites, including the prostaglandinE2, hydroxyeicosatetraenoic acid and the leucotriene B4, have been found to be increased in oral squamous carcinoma. However, the role they play in the potential for malignisation has yet to be studied in detail.

8.20 Enzymes

Glutathione S-transferase (GSTS) is an isoenzyme that acts in the second phase of cell metabolism. It belongs to a complex family of multifunctional proteins and plays an important role in protecting the cell against cytotoxic and carcinogenic agents. There are three types of GST: α, β and π. GST-π is over expressed in human cancer tissue, in premalignant oral lesions and during experimental oral carcinogenesis. Therefore, it may be used as a tumor marker of premalignant oral epithelial lesions. Epithelial dysplasia and GST-π have been found to be related to local immunological dysfunction.

8.21 Molecular Markers of the Risk of Carcinogenesis

8.21.1 Carcinogenesis risk and molecular markers

The clinical appearance and, especially, the degree of dysplasia that may be shown by pre-cancerous lesions in the oral cavity suggest a potential for malignisation. The diagnosis of precancerous lesions begins with the clinical examination it is the histopathological study which determines the presence and degree of any dysplasia (a term covering various alterations of the normal development and maturation of tissue, particularly epithelial. It is postulated that cancer develops as a result of the accumulation of genetic errors in the same tissue — the activation of oncogenes and the inactivation of tumor suppressor genes also being involved. Statistical studies of molecular aspects suggest that between six and ten genetic alterations are required to produce a malignant transformation of the oral mucosa. There are various types of cell and tissue markers that, from a molecular perspective, may provide additional information to that obtained from the clinical examination and histopathological study.

Oral leukoplakia, the most common premalignant lesion of the oral cavity. Although the incidence of oral leukoplakia (white patches) is virtually impossible to ascertain, it surely exceeds the incidence of oral cancer. Leukoplakia is a marker of an increased risk of cancer anywhere in the oral cavity, but there are no reliable clinical or histological features that can be used to predict whether the lesion will regress spontaneously or progress to cancer.

The top panels show the clinical progression of a patient's oral leukoplakia lesion (left-hand panel) to an oral cancer (right-hand panel), which developed three years after the complete resection of the leukoplakia. The middle panels show histologic progression from hyperplasia to cancer (hematoxylin and eosin, x40). The transition from severe dysplasia to an early stage of oral cancer can be seen in the far-right-hand panel. The model of molecular progression shown in the bottom panels of the figure indicates where certain losses of heterozygosity can occur; the cumulative number of genetic alterations is more important than the order in which they occur. A low risk is correlated with no loss of heterozygosity at 3p or 9p; an intermediate risk with loss at 3p, 9p, or both; and a high risk with loss at 3p, 9p, or both plus losses at any of the other chromosomes.

8.22 Survivin

Survivin is a recently described apoptosis inhibitor selectively over-expressed in most tumors. Immunohistochemistry was used to investigate a potential role of survivin as an early predictor of malignant transformation in precancerous and cancerous lesions of the oral cavity. High expression of cytoplasmic/nuclear survivin is an early event during oral carcinogenesis and may provide a useful tool for the identification of precancerous lesions at higher risk of progression into invasive carcinoma.

In 1995 Scully stated that most OSCC arise in the presence of clinical premalignant lesions. In 1999 Schepman *et al.*, reported Erythroplakias and dysplastic leukoplakias are the most frequent potentially malignant lesions, and about half of OSCCs exhibit associated leukoplakia. The presence of severe epithelial dysplasia, morphologically characterized by enlarged nuclei and eosinophilic nucleoli, hyperchromatism, dyskeratosis, and aberrant mitoses, is suggestive of malignant transformation.

Although up to a one third (3–33%) of oral precancerous lesions will eventually evolve into invasive OSCC over a 10-year interval, no reliable histopathological parameters have been identified that predict their potential for subsequent transformation. A prophylactic surgical management is often impractical, especially in patients with multiple and extensive precancerous lesions. Therefore, novel molecular predictors of malignant progression are needed to identify oral precancerous lesions at greater risk of invasive transformation as candidates for surgical intervention.

Survivin is a recently characterized IAP protein which is found abundantly expressed in solid and hematological malignancies, but which is undetectable in most normal adult differentiated tissues. Despite the redundancy of cell death pathways, survivin appears to be required for cancer cell viability, and interference with survivin expression/function has been associated with catastrophic defects of mitotic transition and induction of mitochondrial-induced cell death. Survivin may also provide a reliable indicator of disease progression, and retrospective analysis of various solid tumors has linked survivin expression to decreased overall survival, negative predictive indicators of aggressive disease, resistance to therapy, and accelerated rates of recurrence.

8.23 Table Show Survivin Expression

Survivin is up-regulated early during malignant transformation of the oral cavity, and that its up-regulation is overwhelmingly associated with precancerous lesions that evolved into full-blown invasive squamous cell carcinomas. Consistent with current efforts of genetic fingerprinting of human tumors, novel molecular markers that may predict disease progression, early recurrences, or resistance to therapy are being intensely investigated. Aberrations in apoptotic programs are a hallmark of perhaps all cancers that potentially affect various stages of malignant transformation. The

Table 8.4 Survivin Expression and Statistical Analysis

Groups	No.	Survivin Expression (%)	Mean Survivin Score	Standard Deviation
Normal oral mucosa	5	0 (0)	0	0
Oral epithelial dysplasia without transformation	30	10 (33)	0.63	0.92
Oral epithelial dysplasia with subsequent transformation	16	15 (94)	2.75	1
OSCC	16	16 (100)	2.93	1.12

Table 8.5 Analysis of Survivin Expression in Oral Lesions by One-way Analysis of Variance (ANOVA) with Bonferroni Multiple Comparisons Test

Comparison	Mean Difference	t	P	P value
Oral epithelial dysplasia with subsequent transformation > OSCC	−0.18	0.56	ns	>0.05
Oral epithelial dysplasia with subsequent transformation > oral epithelial dysplasia without transformation	2.11	7.34	* * *	<0.01
Oral epithelial dysplasia with subsequent transformation > normal	2.75	7.33	* * *	<0.01
Oral epithelial dysplasia without transformation > OSCC	−2.30	7.99	* * *	<0.01
Normal > OSCC	−2.93	7.82	* * *	<0.01
Normal > oral epithelial dysplasia without subsequent transformations	−0.63	1.86	ns	>0.05

process of survivin re-expression is an early event during stepwise malignant transformation, which may confer selective growth advantage and resistance to environmental or checkpoint-initiated pro-apoptotic stimuli. The over expression of survivin represents a hallmark of malignant conversion. Naturally, higher scores of survivin expression are likely to be more indicative of this risk. Considering that, in hyperplastic epithelia without associated dysplasia, survivin positivity may reach 5–10% of cells (never exceeding this value), consider that a cut-off value for positivity able to predict malignant conversion.

Al-Rajhi et al., 2000 reported OSCC is a frequent tumor in humans and carries elevated rates of recurrence Mattijssen *et al.*, 1993 states that it may involve up to 80% of patients within 2 yrs. In addition, the five-year survival i.e. in 1998 Friedlander et al., reported rates of OSCC are not encouraging. This stresses the need for new molecular markers of disease progression that could reliably identify patients at high risk of developing invasive disease. In 2001 Altieri, stated that consistent with its predictive/prognostic value in other tumors, histological determination of survivin expression in oral premalignant lesions may provide a quick and potentially useful indicator for identifying patients requiring more aggressive therapeutic intervention.

9

Discussion

Oral carcinogenesis is a multistep process in which multiple genetic events occur that alter the normal functions of oncogenes and tumor suppressor genes. This can result in increased production of growth factors or numbers of cell surface receptors, enhanced intracellular messenger signaling, and/or increased production of transcription factors. In combination with the loss of tumor suppressor activity, this leads to a cell phenotype capable of increased cell proliferation, with loss of cell cohesion, and the ability to infiltrate local tissue and spread to distant sites. Recent advances in the understanding of the molecular control of these various pathways will allow more accurate diagnosis and assessment of prognosis, and might lead the way for more novel approaches to treatment and prevention.

The understanding of the molecular basis of oral squamous carcinoma has increased rapidly over the past few years. Multiple genetic events that culminate in carcinogenesis include the activation of oncogenes and inactivation of tumor suppressor genes. However, not all genetic events occur in all squamous oral carcinomas and similar genetic alterations may occur at different times in the process of carcinogenisis.

As the cellular and molecular biology of cancer advances the therapeutic implications of this molecular biology are also increases as prospective therapy. So various novel treatments including gene transfer, viral vectors, and immune targeting mechanisms can also be provided to the patients.

Oral carcinoma is the most common malignancy of orofacial region though strictly speaking carcinoma means malignancies of epithelial tissue origin.

The malignant neoplasms of epithelial cell origin, derived from any of the three germ layers, are called carcinomas. Thus cancer arising in the epidermis of ectodermal origin is a carcinoma, as is a cancer arising in the mesodermally derived cells of the renal tubules and the endodermally derived cells of the lining of gestrointestinal tract.

Manjul Tiwari (MDS), Tumor Marker & Carcinogenesis, 127–130.
© 2012 *River Publishers. All rights reserved.*

Carcinomas may be further qualified. One with a glandular growth pattern microscopically is termed an adenocarcinoma, and one producing recognizable squamous cells arising in any epithelium of the body is termed a squamous cell carcinoma.

Carcinogenesis is a multistep process at both phenotypic and genetic level means the creation of cancer. It includes the process of derangement of the rate of cell division due to damage to D.N.A. so cancer is ultimately a disease of genes. Cancer is caused by a series of mutations. Each mutation alters the behavior of the cell.

The transformation of normal cells in to malignant cells is dependent on mutations in the genes that control cell cycle progression, leading to the loss of regulatory cell cycle growth signals.

It might then be profitable to list some fundamental principles before we delve in to the details of the molecular basis of cancer.

At the molecular level, progression results from accumulation of genetic lesions that in some instances are favored by defects in D.N.A repair.

Three classes of normal regulatory genes-the growth promoting proto-oncogene, the growth inhibiting cancer suppressor genes (antioncogenes), and genes that regulate programmed cell death, or apoptosis-are the principal targets of genetic damage.

Among the molecular mechanisms involved in the carcinogenesis, defects in the regulation of programmed cell death (apoptosis) may contribute to the pathogenesis and progression of cancer. Dysregulation of oncogenes and tumor suppressor genes involved in apoptosis are also associated with tumor development and progression.

Genes that regulate apoptosis may be dominant, as are proto-oncogene, or they may behave as cancer suppressor genes.

In addition to the three classes of genes mentioned earlier, a fourth category of genes, those that regulate repair of damaged DNA are also pertinent in carcinogenesis.

DNA repair genes affect cell proliferation or survival indirectly by influencing the ability of the organism to repair nonlethal damage in other genes, including proto-oncogenes, tumor suppressor genes, and genes that regulate apoptosis.

The unregulated growth that characterizes cancer is caused by damage to DNA, resulting in mutations to gene that encode for proteins controlling cell division. Many mutation events may be required to transform a normal cell in to a malignant cell. These mutations can be caused by chemicals or physical agents called carcinogens, by close exposure to radioactive materials or by certain viruses that can insert their DNA in to the human genome.

Mutations occur spontaneously, and may be passed down from one generation to the next as a result of mutations within germ line.

The genetic hypothesis of cancer implies that a tumor mass results from the clonal expansion of a single progenitor cell that has incurred the genetic damage. Clonality of tumors is assessed quite readily in women who are heterozygous for polymorphic X-linked markers, such as the enzyme glucose-6-phosphate dehydrogenase (G6PD) or X-linked restriction fragment length polymorphism.

A malignant neoplasm has several phenotypic attributes, such as excessive growth, local invasiveness, and the ability to form distant metastases. These characteristics are acquired in a stepwise fashion, a phenomenon called tumor progression.

Many forms of cancer are associated with exposure to environmental factors such as tobacco smoke, radiation, alcohol and certain viruses.

A variety of agents increase the frequency with which cells are converted to the transformed condition, they are said to be carcinogenic agents. Carcinogens may cause epigenetic changes or may act directly or indirectly to change the genotype of the cells.

Although tobacco is clearly of major aetiological significance (IARC 1984) the failure of overtly malignant lesions to develop in all tobacco users and the development of oral cancer in all tobacco users and the development of oral cancer in persons with no history of tobacco use suggests that the genesis of oral cancer may also involve other unidentified environmental and host factors.

The treatment strategies based on cellular and molecular mechanisms may be divided into conventional or novel categories.

Conventional drug treatments have historically been targeted at DNA replication and cellular proliferation. An exploitation of these modes has led to the development of a new range of therapeutic options by combining conventional agents with others such as calcium channel antagonists.

So carcinogenesis is very important to know the

1. Etiology of cancer in terms of changes in carcinomas at molecular level.
2. Mechanism of development of cancer along with its spread in at all over body including oral cavity.
3. Various carcinogenic agents and their mechanism of action in the development of cancer.

4. Various treatment modalities at the ultra structural level for the benefit of human-beings.

The experimental carcinogenesis provides information about the initiation and progression of oral cancer on the basis of human and animal studies.

Indeed, it is not yet possible to determine with any degree of certainty which lesions displaying the histological feature of dysplasia will inevitably progress to squamous cell carcinoma. It seems likely that in other tissues, is a multifactorial disease in which viral and chemical agents may act synergistically in a host whose immune responses are altered or deficient.

Carcinogenesis means induction of tumors: agents which can induce tumors are called carcinogens. These terms are used for neoplastic proliferation of benign as well as malignant tumors, though the word 'carcino' implies cancer.

The genetic and molecular basis for cancer has long been suspected due to the following observations:

- Hereditary predisposition in some cancers.
- Chromosomal abnormalities in many forms of cancers.
- Correlation of syndrome of defective DNA repair mechanisms and occurrence of cancers.
- Genetic damage (mutagenesis) by the action of various exogenous agents followed by progression to carcinogenesis.

Broadly speaking genes and molecular factors involved in the pathogenesis of cancer can be grouped in to 4 categories-

1. Oncogenes (cancer causing genes)
2. Anti-oncogenes (cancer suppressor genes)
3. Mutator genes (genes that regulate DNA repair)
4. Telomerase in cancer (telomere shortening as cancer suppressor mechanisms)

10

Summary and Conclusion

New tumor markers are identified every year. Only a few have stood the test of time and proved to have clinical usefulness.[48]

The clinical appearance and, especially, the degree of dysplasia that may be shown by pre-cancerous lesions in the oral cavity suggest a potential for malignisation. An increasing number of studies are seeking new, more specific markers that would help to determine the degree of cell alteration and enable a better understanding of the degree of malignant degeneration of these cells.

Clinically it's described as molecules that can be detected in plasma or other body fluids *"Tumor Markers cannot be constructed as primary modalities for the diagnosis of cancer"*.

Biochemical and molecular biological studies that define markers for the evolution of premalignant and malignant lesions will also serve to evaluate prognosis and treatment efficacy. Their main utility in clinical medicine has been as laboratory test to support the diagnosis. Currently, the main use of tumor markers is to asses a cancer's response to treatment and to check for recurrence. Various tissue markers have been identified. One of these is telomerase activity, whose quantification may in the future become a parameter for diagnosing and determining the prognosis of premalignant and malignant oral mucosa lesions.

Many factors are involved in the process of carcinogenesis and it requires the accumulation of multiple genetic alterations in epithelial cells. The inclusion of molecular biology techniques in the pathological diagnosis of biopsies, for both pre-cancerous lesions and squamous cell carcinomas, may improve ostensibly the detection of alterations which are invisible to the microscope. This will enable therapy to be more effective in cases where genetic alterations in the oral mucosa are detected. In the near future it will be possible to identify those patients at high risk of developing oral cancer. A preliminary model of genetic progression has already been reported for

Manjul Tiwari (MDS), Tumor Marker & Carcinogenesis, 131–134.

head and neck squamous cell carcinoma. This model involves detecting the genetic alterations present in premalignant head and neck lesions, along with the genetic regression found in adjacent areas. Certain genetic characteristics of premalignant lesions are thus revealed which, if not treated within a given period, will become aggressive cancerous lesions.

Although the diagnosis of precancerous lesions begins with the clinical examination it is the histopathological study which determines the presence and degree of any dysplasia (a term covering various alterations of the normal development and maturation of tissue, particularly epithelial). It is postulated that cancer develops as a result of the accumulation of genetic errors in the same tissue - the activation of oncogenes and the inactivation of tumor suppressor genes also being involved. But there are various types of cell and tissue markers that, from a molecular perspective, may provide additional information to that obtained from the clinical examination and histopathological study.

Despite extensive research on the biological and molecular aspects of oral SCC, the problems of local-regional recurrence and distant metastasis still persist. Among the more pressing problems in clinical management is the lack of early detection, due to the absence of a potential diagnostic marker. Oncologists are now more aware of the challenges associated with the treatment of cancer of the oral cavity, and survival percentages are improving significantly. More trials are need in the area of improved surgical procedures, variations in dosages of radiotherapy, and the use of various combinations of chemotherapeutic agents with minimal side effects. Moreover, progress in the elucidation of the molecular genetic changes that lead to the development of these tumors should soon bring novel diagnostic and therapeutic procedures into clinical practice.

Because it's important to detect cancer early and to be able to follow it during or after treatment, new tumor markers are still being looked for. But as doctors have learned more about cancer, it's become apparent that the level of a single protein or other substance in the blood may not be the best marker for the disease.

Researchers are starting to focus their attention on genetic markers to detect cancer. We know that most cancers have changes in their DNA, the molecules that direct the functions of all cells. By looking for DNA changes in blood, stool, or urine, scientists may be able to detect cancers very early. The study of patterns of DNA changes (known as *genomics*) is likely to prove more useful than looking for single DNA changes.

It remains to be seen, however, whether the combination of integrin expression with other possible tumor markers is indeed a useful instrument for predicting the malignant transformation of such lesions. The most common genetic alterations in oral cancer are aberrations of thep53 gene, although these alone do not account for its development.

Apoptosis (programmed cell death) in oral carcinoma is lower in poorly differentiated carcinomas but it is the result of loss of FAS expression or increased antiapoptotic factors. Few studies have been undertaken investigate apoptosis in oral cancers, although Jordan and Colleagues demonstrated that Bcl2 was present in differentiated cancers whereas Bax is present in differentiated oral cancers. In this regard it is interesting that recently it is observed in oral primary and metastatic squamous carcinomas an increase expression of Bcl xb, Bax - protein that stimulate apoptosis in contrast to those that inhibit apoptosis including – Bcl -2.

However, the role of this protein in oral cancer is of particular relevance due to its clinical implications.

Another newer approach is called *proteomics*. This technique looks at the pattern of all the proteins in the blood (instead of looking at individual protein levels). New testing equipment allows doctors to look at thousands of proteins at one time. It's unlikely that such a test would be used in a doctor's office, but it may help researchers narrow down which protein levels are important in a particular type of cancer. This information could then be used to develop a blood test that might look only at these important proteins.

These new testing methods are still in the early stages of development. Very few are in routine use at this time.

New developments have found in oral cavity cancers like Abnormal DNA can be found in saliva samples of people with these cancers. It may be a good way to detect them early in people at high risk or who have been treated for these cancers. Research on this approach is under way.

Tumor markers are sometimes elevated in nonmalignant conditions. When a marker is used for cancer screening or diagnosis, the physician must confirm a positive test result by using imaging studies, tissue biopsies, and other procedures. False positive results may occur in laboratory tests when the patient has cross-reacting antibodies that interfere with the test.

Most tumor markers are not useful for screening because levels found in early malignancy overlap the range of levels found in healthy persons. The levels of most tumor markers are elevated in conditions other than malignancy and it is states that measurements of tumor marker levels alone are insufficient to diagnose cancer for the following reasons:

1. Tumor marker levels can be elevated in people with benign conditions.
2. Tumor marker levels are not elevated in every person with cancer, especially in the early stages of the disease.
3. Many tumor markers are not specific to a particular type of cancer.
4. The level of a tumor marker can be elevated by more than one type of cancer.
5. Not every tumor will cause a rise in the level of its associated marker, especially in the early stages of some cancers.

Remarkable advances in several fields of cancer research have enabled a better understanding of the molecular and cellular pathogenesis and progression of cancer of the oral cavity. Those cancer cells which have acquired characteristics conferring a selective growth advantage over their neighbors are successful.

However tumor marker play important role in oral carcinomas like diagnostic, prognostic and therapeutic. It can provide physicians with information used in staging cancers and help in predicting the response to treatment.

In conclusion, research should be continued and extended in all the above mentioned areas, as well as in new ones, so that study of the genome and related factors can be carried out through simpler and cheaper techniques which, in turn, can be applied in routine diagnostic protocols.

References

[1] Abel Sánchez-Aguilera, Margarita Sánchez-Beato, Juan F. García, Ignacio Prieto, Marina Pollan, and Miguel A. Piris; p14^{ARF} nuclear over expression in aggressive B-cell lymphomas is a sensor of malfunction of the common tumor suppressor pathways, Blood, 15 February 2002, Vol. 99, No. 4, pp. 1411–18.

[2] Akihiro Katayama, Nobuyuki Bandoh, Kan Kishibe, Miki Takahara, Takeshi Ogino, Satoshi Nonaka and Yasuaki Harabuchi; Expressions of Matrix Metalloproteinases in Early-Stage Oral Squamous Cell Carcinoma as Predictive Indicators for Tumor Metastases and Prognosis Clinical Cancer Research Vol. 10, 634–40, January 2004.

[3] Anette Gruttgen, Michalea Reichenzeller, Markus Junger, Simone Schilien, Annette Affolter, Franz X. Bosch: Detailed gene expression analysis but not microsatellite marker analysis of 9p 21 reveals differential defects in the INK4a gene locus in the majority of head and neck cancers: *J Oral Pathol* 2001: 194: 311–17.

[4] Bánkfalvi A, Krabort M, Végh A, Felszeghy E, Piffkó J. Deranged expression of the E-cadherin/b-catenin complex and the epidermal growth factor receptor in the clinical evolution and progression of oral squamous cell carcinomas. *J Oral Pathol Med* 2002: 31: 450–7.

[5] Benjamin lewin: genes, oxford university 1997.

[6] Brad W. Neville Oral & maxillofacial pathology, second edition.

[7] BRCA2From Wikipedia, the free encyclopedia.

[8] Cairns P., Polascik T. J., Eby Y., Tokino K., Califano J., Merlo A., Mao L., Herath J., Jenkins R., Westra W., Rutter J. L., Buckler A., Gabrielson E., Tockman M., Cho K. R., Hedrick L., Bova G. S., Isaacs W., Koch W., Schwab D., Sidransky D. Frequency of homozygous deletion at p16/CDKN2 in primary human tumors. *Nat. Genet.* 1995: 11: 210–12.

[9] Califano J, Westra WH, Meininger G, Corio R, Koch WM, Sidransky D. Genetic progression and clonal relationship of recurrent premalignant head and neck lesions. *Clin Cancer Res* 2000: 6: 347–52.

[10] CA-125From Wikipedia, the free encyclopedia.

[11] C.H. Chen, Y.S. Lin, C.C.Lin, Y.H. Yang, Y.P. Ho, C.C. Tsai: Elevated serum levels of a c-erb B -2 oncogene product in oral squamous cell carcinoma patients. *J Oral Pathol Med* 2004: 33: 589–94.

[12] Chung-Ji Liu, Kuo-Wei Chang, Shou-Yee Chao, Po-Cheung Kwan, Shun-Min Chang, Rui-Min Yen, Chun-Yu Wang, Yong-Kie Wong: The molecular markers for prognostic evaluation of areca – associated buccal squamous cell carcinoma. *J Oral Patholmed* 2004: 33: 327–34.

[13] Chung-Ji L, Yann-Jinn L, Hsin-Fu L, Ching-Wen D, Che-Shoa C, Yi-Shing L *et al.* The increase in the frequency of MICA gene A6 allele in oral squamous cell carcinoma. *J Oral Pathol Med* 2002: 31: 323–8.

[14] Cruz I, Napier SS, van der Waal I, Snijders PJ, Walboomers JM, Lamey PJ et al. Suprabasal p53 immunoexpression is strongly associated with high grade dysplasia and risk for malignant transformation in potentially malignant oral lesions from Northern Ireland. *J Clin Pathol* 2002: 55: 98–104.

[15] Das BR et al. – understanding the biology of oral cancer. *Med Sci Monit* 2002; 8(11): RA 258–67.

[16] David Sidranksy, Jay Boyle, Wayne Koch, Peter van der Riet: Oncogene Mutation as intermediate Markers. *J Cellular Biochemistry*, Supplement: 1993, 17F: 184–87.

[17] David T. W. Wong, Peter F. Weller, Stephen J. Galli, Aram Elovic, Thomas H. Rand, Geroge T. Gallagher, Tao Chiang, Ming Yung Chou, Karekine Matossian, Jim Mcbride, Randy Todd. Human Eosinophils Express Transforming Growth Factor Alpha. *J Oral Pathol Med* 1990: 172: 673–81.

[18] De Vicente JC, Sonsoles Olay, Paloma Lequerica-Fernandez, Jacobo Sanchez–Mayoral, Luis Manuel Junquera, Manuel Florentino Freseno. Expression of Bcl-2 but not Bax has a prognostic significance in tongue carcinoma. 2006: 35: 140–5.

[19] Easwar Natarajan, Marcela Saeb, Christopher P. Crum, Sook B. Woo, Phillip H. McKee, and James G. Rheinwald; Co-Expression of p16INK4A and Laminin 5_2 by Microinvasive and Superficial Squamous CellCarcinomas *in Vivo* and by Migrating Wound and Senescent Keratinocytes in Culture. *Am J Pathol* 2003: 163: 477–91.

[20] E M Rosen, S Fan and C Isaacs BRCA1 in hormonal carcinogenesis: basic and clinical research. *Endocrine-Related Cancer* 12 (3) DOI: 10.1677/erc.1.00972 533–48.

[21] Epstein JB, Zhang L, Poh C, Nakamura H, Berean K, Rosin Ml. Increased allelicloss in toluidine blue-positive oral premalignant lesions. *Oral Surg Oral Med Oral Pathol Oral Radiol Endod* 2003: 95: 45–50.

[22] G. Ueda, H. Sunakawa, K. Nakamori, T. Shinya, W. Tsuhako, Y. Tamura, T Kousgi, N. Sato, K. Ogi, H. Hiratsuka: Abeerant expression of beta and alpha catenin is an independent prognostic marker in oral squamous cell carcinoma. *Int J Oral Maxillofac Surg* 2006: 35: 356–61.

[23] H. Myoung, M.J. Kim, J.H lee, Y.J Ok, J.Y. Paeng, P.Y. Yun; Correlation of proliferative markers (Ki-67 and PCNA) with survival and lymph node metastasis in oral squamous cell carcinoma: a clinical and histopathological analysis of 113 patients. *Int J Oral Maxillofac Surg* 2006: 35: 1005–10.

[24] H K Williams. Molecular pathogenesis of oral squamous carcinoma. *J Clin Pathol: Mol Pathol* 2000: 53: 165–72.

[25] Hirohumi Arakawa, Feng Wu, Max Costa, William Rom, Moon–shong Tang; Sequence specificity of Cr (III)–DNA adduct formation in the p^{53} gene: NGG sequence are preferential adduct–forming sites. 2006: 27: 3: 639–45.

[26] Hiroyuki Suzuki, Haruhiko Sugimura, Kenji Hashimoto: $p^{16INK4A}$ in oral squamous cell carcinoma – a correlation with biological behaviors: immunohistohostochemical and FISH analysis. *J Oral Maxillofac Surg* 2006: 64: 1617–23.

[27] http://www.cancer.org/docroot/ipg.asp?Site name=National+cancer+INSTITUTE& URL=HTTP://WWW.CANCER.ORG

[28] http://www.cechtuma.cz/bioenv/1997/2/c122-en.html

[29] Jesper Reibel Prognosis of oral pre-malignant lesions: significance of clinical, histopathological, and molecular biological characteristics. *Crit Rev Oral Biol Med* 2003: 14(1): 47–62.

[30] Jurgen Behrens, Luc Vakaet, Roberrt Friis, Elke Winterhager, Frans Van Roy, Marc M. Mareel, Walter Birchmeier: Loss of epithelial differentiation and gain of invasiveness correlates with Tyrosine Phos phorlytaion of the E–Cadhrein/beta Catenin in cells transformed with a temperature-sensitive v–SRC gene: *J Cell Biology* 1993: 120: 3: 757–66.

[31] K Park, park's textbook of preventive and social medicine 17th edition.

[32] Karin Nylander, Erik Dabelsteen, Peter A. Hall; The p[53] molecule and its prognostic role in squamous cell carcinoma of head and neck. *J Oral Pathol Med* 2000: 29: 413–25.

[33] Kuang–Yu Jen, Vivian G. Cheung. Identification of Novel p[53] Target Genes in Ionizing Radiation Response. *J Cancer Res* 2005: 65(17): 7666–73.

[34] Lucimari Bizari[I]; Ana Elizabete Silva[I]; Eloiza H. Tajara[II], Gene amplification in carcinogenesis[I] Universidade Estadual Paulista, Departamento de Biologia, São José do Rio Preto, SP, Brazil[II] Faculdade de Medicina de São José do Rio Preto, Departamento de Biologia Molecular, São José do Rio Preto, SP, Brazil (Genet. Mol. Biol. vol. 29 no. 1 São Paulo 2006).

[35] Lumerman H., Freedman P., Kerpel S. Oral epithelial dysplasia and the development of invasive squamous carcinoma. *Oral Surg. Oral Med. Oral Pathol. Oral Radiol. Endod.* 1995: 79: 321–29.

[36] Mei-Ling Kuo, Eric J. Duncavage, Rose Mathew, Willem den Besten, Deqing Pei, Deanna Naeve, Tadashi Yamamoto, Cheng Cheng, Charles J. Sherr and Martine F. Roussel. Arf Induces p53-dependent and -independent Antiproliferative Genes [*Cancer Research* 63, 1046–53, March 1, 2003]

[37] Mineta H, Miura K, Takebayashi S, Ueda Y, Misawa K, Haida H et al. Cyclin D1 overexpression correlates with poor prognosis in patients with tongue squamous cell carcinoma. *Oral Oncol* 2000: 36: 194–8.

[38] MYC From Wikipedia, the free encyclopedia.

[39] Noriyuki Akita, Fukuto Maruta, Leonard W. Seymour, David J. Kerr, Alan L. Parker, Tomohiro Asai, Naoto Oku, Jun Nakayama, Shinici Miyagawa: dentification og oligopeptides bindin to peritoneal tumors of gastric cancer. *J Cancer Sci* 2006: 97: 10: 1075–81.

[40] OncogeneFrom Wikipedia, the free encyclopedia.

[41] P.J. Thomson, M.L. Goodson. C. Booth, N. Cragg: Cyclin A activity predicts clinical outcome in oral precancer and cancer. *Int. J. Oral Maxiloofac. Surg.* 2206: 35: 1041–46.

[42] P. Sdek, Z.Y. Zhang, J. Cao, H.Y. Pan, W.T. Chen, J.W. Zheng: Alteration of cell cycle regulatory proteins in human oral epithelial cells immortalized by HPV 16 E6 and E7. *Int J Oral Maxillofac* 2006: 35: 653–57.

[43] Patricia A. Hoffee, Fence creek, Madison, Connecticut; Genetics.

[44] Piattelli A, Rubini C, Fioroni M, Iezzi G, Santinelli A. Prevalence of p53, bcl-2 andki-67 Immunoreactivity and of apoptosis in normal epithelium and in premalignant and malignant lesions of the oral cavity. J Oral Maxillofac Surg 2002: 60: 532–40.

[45] Reed A. L., Califano J., Cairns P., Westra W. H., Jones R. M., Koch W., Ahrendt S., Eby Y., Sewell D., Nawroz H., Bartek J., Sidransky D. High frequency of *p16*

(*CDKN2/MTS-1/INK4A*) inactivation in head and neck squamous cell carcinoma. *Cancer Res.* 1996: 56: 3630–33.

[46] Rhonda A. Kwong, Larry H. Kalish, Tuan V. Nguyen, James G. Kench, Ronaldo J. Bova, Ian E. Cole, Elizabeth A. Musgrove, and Robert L. Sutherland p14 ARF Protein Expression Is a Predictor of Both Relapse and Survival in Squamous Cell Carcinoma of the Anterior Tongue.

[47] Ricardo D. Coleta, Paola Cotrim, Pablo Agustin Vargas, Halbert Villalba, Fabio Ramoa Pires, Marcio de Morases, Oslei Paes de Almeida, Piracicababa–SP: Basoloid squq-mous carcinoma of the oral cavity : Report of 2 cases and study of AgNOR, PCNA, p[53] and MMp Expression. *Oral Surg Oral Med Oral Pathol Oral Radiol Endod* 2001: 91: 539–563.

[48] Robbins and Cotran Pathologic Basis of Disease, 7TH Edition.

[49] S.B Pai, R.B. Pai, R.M. Lalitha, S.V. Kumarswamy, N. Lalitha, R.N. Johnston, M.K. Bhargava: Expression of oncofoetal marker carcinoembryonic antigen in oral cancers in South India – a pilot study. *Int J Oral Maxillofac Surg* 2006: 35: 746–49.

[50] S. K. Mohanty and P. Dey. Serous effusions: diagnosis of malignancy beyond cytomor-phology. An analytic review Postgraduate Medical Journal 2003: 79: 569–74.

[51] Schliephake H. Prognostic relevance of molecular markers of oral cancer-A review. *J Oral Maxillofac Surg* 2003: 32: 233–45.

[52] Seiji Hoska, Tetsuya Nakastra, Hirotake Tsukamota, Takumi Hatayama, Hideo Baba, Yasuharu Nishimura. Synthetic Small interfering RNA targeting heat shock protein 105 induces apoptosis of various cancer cell both in vivo and in vitro. *J Cance Sci* 2006: 97: 7: 623–32.

[53] Shan Gao, Hans Eiberg, Annelise Krogdahl, Chung-Ji L. iu, Jens Ahm Sorensen: Cyto-plasmic expression of E-cadherin and beta cateinin correlated woth LoH and hyper methylation of the APC gene in oral squamous cell carcinoma. *J Oral Pathol Med* 2005: 34: 116–19.

[54] Shao-Chen Liu, Edward R. Sauter, Margie L. Clapper, Roy S. Feldman, Lawrence Levin, Sow-Yeh Chen, Timothy J. Yen, Eric Ross, Paul F. Engstrom, and Andres J. P. Klein-Szanto 72 Markers of Cell Proliferation in Normal Epithelia and Dysplastic Leukoplakias of the Oral Cavity Vol. 7, 597–603, Juls 1998.

[55] Shilpi Arora, Jatinder Kaur, Chavvi Sharma, Meera Mathur, Sudhir Bahadur, Nootan K. Shukla, Suryanaryana V.S. Deo and Ranju Ralhan1. Stromelysin, Ets-1, and Vascular Endothelial Growth Factor Expression in Oral Precancerous and Cancerous Lesions: Correlation with Micro vessel Density, Progression, and Prognosis Clinical Cancer Research Vol. 11, 2272–84, March 2005.

[56] Sturat P. Atkinson, Stacey F. Hoare, Rosalind M. Glasspool, W. Nicol Keith: Lack of Telomerase Gene Expression in Alternative Lengthening of Telomere Cells is associ-ated with chromatin remodeling of the Htr and Htert Gene Promoters. *J Cancer Res* 2005: 65(17) 7585–90.

[57] Sudbo J, Bryne M, Johannessen AC, Kildal W, Danielsen HE, Reith A. Comparison of histological grading and large-scale genomic status (DNA ploidy) as prognostic tools in oral dysplasia. *J Pathol* 2001: 194: 303–13.

[58] T. Fillies, H. Buerger, C. Gaertner, C. August, B. Brandt. U. Joos, R. Werkmeister: Catenin expression in T1/2 carcinomas of the floor of the mouth. *Int J Oral Maxillofac Surg* 2005: 34: 907–11.

[59] Thomas Fillies, Richard Werkmeister, Jens Packeisen, Burkhard Brandt, Philippe Morin5, Dieter Weingart, Ulrich Joos and Horst Buerger Cytokeratin 8/18 expression indicates a poor prognosis in squamouscell carcinomas of the oral cavity *BMC Cancer* 2006, 6: 10 doi:10.1186/1471-2407-6-10.

[60] Torsten E. Reichert, M.D., Ph.D., Claudia Scheuer, Ph.D., Roger Day, Sc. D. Wilfried Wagner, M.D., Ph.D., Theresa L. Whiteside, Ph.D., The Number of Intratumoral Dendritic Cells and z-Chain Expression in T Cells as Prognostic and Survival Biomarkers in Patients with Oral Carcinoma Presented at the 5th International Conference for Head and Neck Cancer, San Francisco, California, July 29–August 2, 2000.

[61] Torsten E. Reichert, Roger Day, Eva M. Wagner, and Theresa L. Whiteside. Absent or Low Expression of the Â Chain in T Cells at the Tumor Site Correlates with Poor Survival in Patients with Oral Carcinoma [CANCER RESEARCH 58. 5344–47. December 1. 1998].

[62] Tsuneo Kobayashi, M.D., and Tomoko Kawakubo, M.Sc., Prospective Investigation of Tumor Markers and Risk Assessment in Early Cancer Screening.

[63] Tumor marker test: www.Google.com: 9.

[64] Wakulich C, Jackson-Boeters L, Daley TD, Wysocki GP. Immunohistochemical localization of growth factors fibroblast growth factor-1 and fibroblast growth factor-2and receptors fibroblast growth factor receptor-2 and fibroblast growth factor receptor-3in normal oral epithelium, epithelial dysplasias, and squamous cell carcinoma. *Oral SurgOral Med Oral Pathol Oral Radiol Endod* 2002: 93: 573–9.

[65] Waun K. Hong, and Reuben Lotan. Increased Expression of Cytokeratins CK8 and CK19 Is Associated with Head and Neck Vol. 4. 87/-8'76. DECEMBER 1995.

[66] Werkmeister R, Brandt B, Joos V. Clinical relevance of erbB-1 and -2 oncogenes in oral carcinomas. Oral Oncol 2000: 36: 100–5.

[67] Yanamoto S, Kawasaki G, Yoshitomi Di, Mizuno A. P53, mdm2 and p21 expression in oral squamous cell carcinomas: Relationship with clinicopathologic factors. *Oral Surg Oral Med Oral Pathol Oral Radiol Endod* 2002: 94: 593–600.

[68] Yasusei Kudo, Shojiro Kitajima, Ikuko Ogawa, Masae Kitagawa, Mutsumi Miyauchi, and Takashi Takata. Small interfering RNA targeting of Sphasekinase Interacting protein 2 inhibits cell growth of oral cancer cells by inhibiting, p. 27 degradation [*Mol Cancer Ther* 2005: 4(3): 471–6].

[69] Yasuyuki Nakamura, Manabu Futamura, Hirok Kamino, Koji Yoshida, Yusuke Nakamura, Hirofumi Arakawa. Identification of p^{53-46f} asa super p^{53} with an enhanced ability to induce p^{53} dependent apoptosis. *J Cancer Sci* 2006: 97: 7: 633–41.

[70] Behrens J, Vakaet L, Frii R, et al. Loss of epithelial differentiation and gain of invasiveness correlates with tyrosine phosphorylation of the E-cadherin/ßcatenin complex in cells transformed with a temperature-sensitive V-SRC gene. *J Cell Biol* 1993: 120: 757–66.

[71] Brachmann RK, Vidal M, Boeke JD. Dominant-negative mutations in yeast hit cancer hot spots. *Proc Natl Acad Sci USA* 1996: 93: 4091–5.

[72] Brennan A, Boyle JO, Koch WM. Association between cigarette smoking and mutation of the p53 gene in squamous cell carcinoma of the head and neck. *N Engl J Med* 1995: 332: 712–17.

[73] Datto MB, Li Y, Panus JF, et al. Transforming growth factor induces the cyclin-dependent kinase inhibitor p21 through a p53-independent mechanism. *Proc Natl Acad Sci USA* 1995: 92: 5545–9.

[74] Dixon K, Kopras E (2004). "Genetic alterations and DNA repair in human carcinogenesis." *Semin Cancer Biol* 14 (6): 441–8.

[75] Dowell SP, Hall PA. The p53 tumor suppressor gene and tumor prognosis. is there a relationship? *Pathol* 1995: 177: 221–24.

[76] E M Rosen, S Fan and C Isaacs BRCA1 in hormonal carcinogenesis: basic and clinical research. 533–48.

[77] Fearon ER, Vogelstein B (1990). "A genetic model for colorectal tumorigenesis". *Cell* 61 (5): 759–67.

[78] Folkman J, Klagsbrun M. Angiogenic factors. *Science* 1987: 235: 442–9.

[79] Hae-Ryun Kim, Russell Christensen, Noh-Hyun Park, Philip Sapp, Mo K. Kang and No-Hee. Park Elevated Expression of *hTERT* Is Associated with Dysplastic Cell Transformation during Human Oral Carcinogenesis *in Situ* Clinical Cancer Research Vol. 7, 3079–86, October 2001.

[80] Harsh Mohan. Essential pathology for dental students second edition 221–32.

[81] H K Williams molecular pathogenesis of oral squamous carcinoma journal of clinical pathology 2000, 53: 165–72.

[82] Hollstein M, Sidransky D, Volgelstein B, et al. p53 mutations in human cancers. *Science* 1991: 253: 49–53.

[83] Dr. Irma B. Gimenez-Conti, DDS, PhD*, Thomas J. Slaga, PhD. The hamster cheek pouch carcinogenesis model, *Journal of Cellular Biochemistry* Vol. 53, Issue S17F, Pages 83–90.

[84] Korineck V, Barker N, Morin PJ, et al. Constitutive transcriptional activation by a ß-catenin–Tcf complex in APC–/–colon carcinoma. *Science* 1997: 275: 1784–7.

[85] Kumar cotran Robbins pathologic basis of disease sixth edition 145–74.

[86] Lee WH. Tumor suppressor genes—the hope [editorial]. *FASEB J* 1993: 7: 819.

[87] Lianes P, Orlow I, Zhang ZF, et al. Altered patterns of mdm2 and TP53 expression in human bladder cancer. *J Natl Cancer Inst* 1994: 86: 1325–30.

[88] Li-Kuo S, Vogelstein B, Kinzler KW. Association of the APC tumor suppressor protein with catenins. *Science* 1993: 262: 1734–7.

[89] Liao PH, Chang YC, Huang MF, Tai KW, Chou MY. Mutation of p53 gene codon 63 in saliva as a molecular marker for oral squamous cell carcinoma.

[90] Lucimari Bizari, Ana Elizabete Silva; Eloiza H. Tajara Gene amplification in carcinogenesis Genet. Mol. Biol. vol. 29 no. 1 São Paulo 2006.

[91] Morin PJ, Sparks AB, Korinek V, et al. Activation of ß-catenin—Tcf signalling in colon cancer by mutations in ß-catenin or APC. *Science* 1997: 275: 1787–90.

[92] Murry W. Hill and Christopher A. Squier Experimental aspects of oral carcinogenesis 492–501.

[93] Ravi meharotra, Anurag gupta, mamta singh and Rahela Ibrahim application of cytology and molecular biology in diagnosing premalignant or malignant oral lesions.

[94] Sarasin A (2003). "An overview of the mechanisms of mutagenesis and carcinogenesis." *Mutat Res* 544 (2–3): 99–106. PMI.

[95] Schottenfeld D, Beebe-Dimmer JL (2005). "Advances in cancer epidemiology: understanding causal mechanisms and the evidence Knudson AG (2001). "Two genetic hits (more or less) to cancer". *Nat Rev Cancer* 1 (2): 157–62.

[96] Shah, Johnson Batskin book of oral cancer. 167–80.

[97] Shiohara M, El-Deiry WS, Wada N, et al. Absence of WAF1 mutations in a variety of human malignancies. *Blood* 1994: 84: 3781–4.

[98] Todd R, Chou MY, Matossian K, et al. Cellular sources of transforming growth factor-alpha in human oral cancer. *J Dent Res* 1990: 70: 917–23.

[99] Vokes EE, Weichselbaum RR, Lippman SM, et al. Head and neck cancer. *N Engl J Med* 1993: 328: 184–94.

[100] Wong DT, Weller PF, Galli SJ, et al. Human eosinophils express transforming growth factor alpha. *J Exp Med* 1990: 172: 673–81.

[101] Yoshida T, Miyagawa K, Odagiri H, et al. Genomic sequence of hst, a transforming gene encoding a protein homologous to fibroblast growth factor and the int-2 encoded protein. *Proc Natl Acad Sci USA* 1987: 84: 7305–9.

[102] Abel Sánchez-Aguilera, Margarita Sánchez-Beato, Juan F. García, Ignacio Prieto, Marina Pollan, and Miguel A. Piris; p14^{ARF} nuclear over expression in aggressive B-cell lymphomas is a sensor of malfunction of the common tumor suppressor pathways, Blood, 15 February 2002, Vol. 99, No. 4, pp. 1411–1418

[103] Akihiro Katayama, Nobuyuki Bandoh, Kan Kishibe, Miki Takahara, Takeshi Ogino, Satoshi Nonaka and Yasuaki Harabuchi; Expressions of Matrix Metalloproteinases in Early-Stage Oral Squamous Cell Carcinoma as Predictive Indicators for Tumor Metastases and Prognosis Clinical Cancer Research Vol. 10, 634–640, January 2004

[104] Anette Gruttgen, Michalea Reichenzeller, Markus Junger, Simone Schilien, Annette Affolter, Franz X. Bosch: Detailed gene expression analysis but not microsatellite marker analysis of 9p 21 reveals differential defects in the INK4a gene locus in the majority of head and neck cancers: *J Oral Pathol*: 2001: 194: 311–317

[105] Bánkfalvi A, Krabort M, Végh A, Felszeghy E, Piffkó J. Deranged expression of the E-cadherin/b-catenin complex and the epidermal growth factor receptor in the clinical evolution and progression of oral squamous cell carcinomas. *J Oral Pathol Med* 2002; 31: 450–7.

[106] Benjamin lewin: genes, oxford university 1997

[107] Brad W. Neville Oral & maxillofacial pathology, second edition

[108] BRCA2 From Wikipedia, the free encyclopedia

[109] Cairns P., Polascik T. J., Eby Y., Tokino K., Califano J., Merlo A., Mao L., Herath J., Jenkins R., Westra W., Rutter J. L., Buckler A., Gabrielson E., Tockman M., Cho K. R., Hedrick L., Bova G. S., Isaacs W., Koch W., Schwab D., Sidransky D. Frequency of homozygous deletion at p16/CDKN2 in primary human tumors. *Nat. Genet.*, 11: 210–212, 1995.

[110] Califano J, Westra WH, Meininger G, Corio R, Koch WM, Sidransky D. Genetic progression and clonal relationship of recurrent premalignant head and neck lesions. *Clin Cancer Res* 2000; 6: 347–52.

[111] CA-125 From Wikipedia, the free encyclopedia

[112] C.H. Chen, Y.S. Lin . C.C.Lin, Y.H.Yang, Y.P.Ho, C.C. Tsai: Elevated serum levels of a c-erb B -2 oncogene product in oral squamous cell carcinoma patients: *J Oral Pathol Med*: 2004: 33: 589–594

[113] Chung-Ji Liu, Kuo-Wei Chang, Shou-Yee Chao, Po-Cheung Kwan, Shun-Min Chang, Rui-Min Yen, Chun-Yu Wang, Yong-Kie Wong: The molecular markers for prognostic evaluation of areca – associated buccal squamous cell carcinoma: *J Oral Pathol med*: 2004: 33: 327–334

[114] Chung-Ji L, Yann-Jinn L, Hsin-Fu L, Ching-Wen D, Che-Shoa C, Yi-Shing L *et al.* The increase in the frequency of MICA gene A6 allele in oral squamous cell carcinoma. *J Oral Pathol Med* 2002; 31: 323–8

[115] Cruz I, Napier SS, van der Waal I, Snijders PJ, Walboomers JM, Lamey PJ *et al.* Suprabasal p53 immunoexpression is strongly associated with high grade dysplasia and risk for malignant transformation in potentially malignant oral lesions from Northern Ireland. *J Clin Pathol* 2002; 55: 98–104

[116] Das BR *et al.*–understanding the biology of oral cancer; *Med Sci Monit*, 2002; 8(11): RA258–267

[117] David Sidranksy, Jay Boyle, Wayne Koch, Peter van der Riet: Oncogene Mutation as intermediate Markers: *J Cellular Biochemistry*, Supplement: 1993, 17F: 184–187

[118] David T. W. Wong, Peter F. Weller, Stephen J. Galli, Aram Elovic, Thomas H.Rand, Geroge T. Gallagher, Tao Chiang, Ming Yung Chou, Karekine Matossian, Jim Mcbride, Randy Todd; Human Eosinophils Express Transforming Growth Factor Alpha; *J Oral Pathol Med*; 1990; 172; 673–681

[119] De Vicente JC, Sonsoles Olay, Paloma Lequerica–Fernandez, Jacobo Sanchez–Mayoral, Luis Manuel Junquera, Manuel Florentino Freseno; Expression of Bcl-2 but not Bax has a prognostic significance in tongue carcinoma; 2006; 35; 140–5

[120] Easwar Natarajan, Marcela Saeb, Christopher P. Crum, Sook B. Woo, Phillip H. McKee, and James G. Rheinwald; Co-Expression of p16INK4A and Laminin 5_2 by Microinvasive and Superficial Squamous CellCarcinomas *in Vivo* and by Migrating Wound and Senescent Keratinocytes in Culture *Am J Pathol* 2003, 163: 477–491

[121] E M Rosen, S Fan and C Isaacs, BRCA1 in hormonal carcinogenesis: basic and clinical research Endocrine-Related Cancer 12(3) DOI: 10.1677/erc.1.00972 533–548

[122] Epstein JB, Zhang L, Poh C, Nakamura H, Berean K, Rosin Ml. Increased allelicloss in toluidine blue-positive oral premalignant lesions. *Oral Surg Oral Med Oral Pathol Oral Radiol Endod* 2003; 95: 45–50.

[123] G. Ueda, H. Sunakawa, K. Nakamori, T.Shinya, W. Tsuhako, Y. Tamura, T Kousgi, N . Sato, K. Ogi, H. Hiratsuka: Abeerant expression of beta and alpha catenin is an independent prognostic marker in oral squamous cell carcinoma: *Int J Oral Maxillofac Surg*: 2006: 35: 356–361

[124] H. Myoung, M.j. kim, J.H lee, Y.J Ok, J.Y. Paeng, P.Y. Yun; Correlation of proliferative markers (Ki-67 and PCNA) with survival and lymph node metastasis in oral squamous cell carcinoma: a clinical and histopathological analysis of 113 patients; *Int J Oral Maxillofac Surg*: 2006: 35: 1005–1010

[125] H K Williams Molecular pathogenesis of oral squamous carcinoma *J Clin Pathol: Mol Pathol* 2000; 53: 165–172

[126] Hirohumi Arakawa, Feng Wu, Max Costa, William Rom, Moon–shong Tang; Sequence specificity of Cr (III)–DNA adduct formation in the p53 gene: NGG sequence are preferential adduct–forming sites; 2006; 27: 3: 639–645

[127] Hiroyuki Suzuki, Haruhiko Sugimura, Kenji Hashimoto: p. 16 INK4A in oral squamous cell carcinoma–a correlation with biological behaviors: immuno histohosto chemical and FISH analysis: *J Oral Maxillofac Surg*: 2006: 64: 1617–1623

[128] http://www.cancer.org/docroot/ipg.asp?Sitename=National+cancerINSTITUTE& URL=HTTP://WWW.CANCER.ORG

[129] http://www.cechtuma.cz/bioenv/1997/2/c122-en.html

[130] Jesper Reibel Prognosis of oral pre-malignant lesions: significance of clinical, histopathological, and molecular biological characteristics 14(1): 47–62 (2003 *Crit Rev Oral Biol Med*

[131] Jurgen Behrens, Luc Vakaet, Roberrt Friis, Elke Winterhager, Frans Van Roy, Marc M.Mareel, Walter Birchmeier: Loss of epithelial differentiation and gain of invasiveness correlates with Tyrosine Phos phorlytaion of the E–Cadhrein/beta Catenin in cells transformed with a temperature-sensitive v–SRC gene: *J Cell Biology*: 1993: 120: 3: 757–766

[132] K Park, Park's textbook of preventive and social medicine 17 edition

[133] Karin Nylander, Erik Dabelsteen, Peter A. Hall; The p53 molecule and its prognostic role in squamous cell carcinoma of head and neck; *J Oral Pathol Med* 2000; 29; 413–425

[134] Kuang–Yu Jen, Vivian G. Cheung: Identification of Novel p53 Target Genes in Ionizing Radiation Response: *J Cancer Res*: 2005: 65(17) 7666–73

[135] Lucimari Bizari I Ana Elizabete Silva; Eloiza H. Tajara, Gene amplification in carcinogenesis I Universidade Estadual Paulista, Departamento de Biologia, São José do Rio Preto, SP, Brazil II Faculdade de Medicina de São *José do Rio Preto*, Departamento de Biologia Molecular, São José do Rio Preto, SP, Brazil (*Genet. Mol. Biol.* vol. 29 no. 1 São Paulo 2006)

[136] Lumerman H., Freedman P., Kerpel S. Oral epithelial dysplasia and the development of invasive squamous carcinoma. *Oral Surg. Oral Med. Oral Pathol. Oral Radiol. Endod.*, 79: 321–329, 1995.

[137] Mei-Ling Kuo, Eric J. Duncavage, Rose Mathew, Willem den Besten, Deqing Pei, Deanna Naeve, Tadashi Yamamoto, Cheng Cheng, Charles J. Sherr and Martine F. Roussel; Arf Induces p53-dependent and independent Antiproliferative Genes Cancer Research 63, 1046–1053, March 1, 2003

[138] Mineta H, Miura K, Takebayashi S, Ueda Y, Misawa K, Haida H *et al.* Cyclin D1 over expression correlates with poor prognosis in patients with tongue squamous cell carcinoma. *Oral Oncol* 2000; 36: 194–8.

[139] MYC From Wikipedia, the free encyclopedia

[140] Noriyuki Akita, Fukuto Maruta, Leonard W. Seymour, David J. Kerr, Alan L. Parker, Tomohiro Asai, Naoto Oku, Jun Nakayama, Shinici Miyagawa: dentification of oligopeptides binding to peritoneal tumors of gastric cancer: *J Cancer Sci*: 2006: 97: 10: 1075–1081

[141] OncogeneFrom Wikipedia, the free encyclopedia

[142] P.J. Thomson,,M.L. Goodson, C. Booth, N. Cragg: Cyclin A activity predicts clinical outcome in oral precancer and cancer. *Int. J. Oral Maxiloofac. Surg.* 2206; 35: 1041–1046

[143] P. Sdek, Z.Y. Zhang, J. Cao, H.Y. Pan, W.T. Chen, J.W . Zheng: Alteration of cell cycle regulatory proteins in human oral epithelial cells immortalized by HPV 16 E6 and E7: *Int J Oral Maxillofac*: 2006: 35: 653–657

[144] Patricia A. Hoffee, Fence creek, Madison, Connecticut; Genetics

[145] Piattelli A, Rubini C, Fioroni M, Iezzi G, Santinelli A. Prevalence of p53, bcl-2 andki-67 Immunoreactivity and of apoptosis in normal epithelium and in premalignant and malignant lesions of the oral cavity. *J Oral Maxillofac Surg* 2002; 60: 532–40.

[146] Reed A. L., Califano J., Cairns P., Westra W. H., Jones R. M., Koch W., Ahrendt S., Eby Y., Sewell D., Nawroz H., Bartek J., Sidransky D. High frequency of p16 (CDKN2/MTS-1/INK4A) inactivation in head and neck squamous cell carcinoma. *Cancer Res.*, 56: 3630–3633, 1996.

[147] Rhonda A. Kwong, Larry H. Kalish, Tuan V. Nguyen, James G. Kench, Ronaldo J. Bova, Ian E. Cole, Elizabeth A. Musgrove, and Robert L. Sutherland p14ARF Protein Expression Is a Predictor of Both Relapse and Survival in Squamous Cell Carcinoma of the Anterior Tongue

[148] Ricardo D. Coleta, Paola Cotrim, Pablo Agustin Vargas, Halbert Villalba, Fabio Ramoa Pires, Marcio de Morases, Oslei Paes de Almeida, Piracicababa–SP: Basoloid squqmous carcinoma of the oral cavity: Report of 2 cases and study of AgNOR, PCNA, p53 and MMp Expression; *Oral Surg Oral Med Oral Pathol Oral Radiol Endod*: 2001; 91: 563–539

[149] Robbins and Cotran Pathologic Basis of Disease, 7TH Edition

[150] S.B Pai, R.B. Pai, R.M. Lalitha, S.V. Kumarswamy, N. Lalitha, R.N. Johnston, M.K. Bhargava: Expression of oncofoetal marker carcinoembryonic antigen in oral cancers in South India–a pilot study: *Int J Oral Maxillofac Surg*: 2006: 35: 746–749

[151] S K Mohanty and P Dey Serous effusions: diagnosis of malignancy beyond cytomorphology. An analytic review Postgraduate Medical Journal 2003; 79: 569–574

[152] Schliephake H. Prognostic relevance of molecular markers of oral cancer—A review. *J Oral Maxillofac Surg* 2003; 32: 233–45.

[153] Seiji Hoska, Tetsuya Nakastra, Hirotake Tsukamota, Takumi Hatayama, Hideo Baba, Yasuharu Nishimura: Synthetic Small interfering RNA targeting heat shock protein 105 induces apoptosis of various cancer cell both *in vivo* and *in vitro*; *J Cance Sci*: 2006: 97: 7: 623–632

[154] Shan Gao, Hans Eiberg, Annelise Krogdahl, Chung–Ji L iu, Jens Ahm Sorensen: Cytoplasmic expression of E-cadherin and beta cateinin correlated woth LoH and hyper methylation of the APC gene in oral squamous cell carcinoma: *J Oral Pathol Med*: 2005: 34: 116–119

[155] 'Shao-Chen Liu, Edward R. Sauter, Margie L. Clapper, Roy S. Feldman, Lawrence Levin, Sow-Yeh Chen,Timothy J. Yen, Eric Ross, Paul F. Engstrom, andAndres J. P. Klein-Szanto 72 Markers of Cell Proliferation in Normal Epithelia and Dysplastic Leukoplakias of the Oral Cavity Vol. 7, 597–603, July 1998

[156] Shilpi Arora, Jatinder Kaur, Chavvi Sharma, Meera Mathur, Sudhir Bahadur, Nootan K. Shukla, Suryanaryana V.S. Deo and Ranju Ralhan1 Stromelysin, Ets-1, and Vascular Endothelial Growth Factor Expression in Oral Precancerous and Cancerous Lesions: Correlation with Micro vessel Density, Progression, and Prognosis Clinical Cancer Research Vol. 11, 2272–2284, March 2005

[157] Sturat P. Atkinson, Stacey F. Hoare, Rosalind M. Glasspool, W. Nicol Keith: Lack of Telomerase Gene Expression in Alternative Lengthening of Telomere Cells is associated with chromatin remodeling of the Htr and Htert Gene Promoters: *J Cancer Res*: 2005: 65(17) 7585–90

[158] Sudbo J, Bryne M, Johannessen AC, Kildal W, Danielsen HE, Reith A. Comparison of histological grading and large-scale genomic status (DNA ploidy) as prognostic tools in oral dysplasia. *J Pathol* 2001; 194: 303–13

[159] T. Fillies, H. Buerger, C. Gaertner, C. August, B. Brandt, U. Joos, R. Werkmeister; Catenin expression in T1/2 carcinomas of the floor of the mouth; *Int J Oral Maxillofac Surg*: 2005, 34: 907–911

[160] Thomas Fillies, Richard Werkmeister, Jens Packeisen, Burkhard Brandt, Philippe Morin5, Dieter Weingart, Ulrich Joos and Horst Buerger Cytokeratin 8/18 expression indicates a poor prognosis in squamous cell carcinomas of the oral cavity BMC Cancer 2006, 6:10 doi:10.1186/1471–2407-6–10

[161] Torsten E. Reichert, M.D., Ph.D.,Claudia Scheuer, Ph.D., Roger Day, Sc.D. Wilfried Wagner, M.D., Ph.D., Theresa L. Whiteside, Ph.D., The Number of Intratumoral Dendritic Cells and-Chain Expression in T Cells as Prognostic andSurvival Biomarkers in Patients with Oral Carcinoma Presented at the 5th International Conference for Head and Neck Cancer, San Francisco, California, July 29–August 2, 2000.

[162] Torsten E. Reichert, Roger Day, Eva M. Wagner, and Theresa L. Whiteside Absent or Low Expression of the Â£Chain in T Cells at the Tumor Site Correlates with Poor Survival in Patients with Oral Carcinoma [Cancer Research 58, 5344–5347, December 1 1998]

[163] Tsuneo Kobayashi, M.D., and Tomoko Kawakubo, M.Sc., Prospective Investigation of Tumor Markers and Risk Assessment in Early Cancer Screening

[164] Tumor marker test: www.Google.com:9

[165] Wakulich C, Jackson-Boeters L, Daley TD, Wysocki GP. Immunohistochemical localization of growth factors fibroblast growth factor-1 and fibroblast growth factor-2and receptors fibroblast growth factor receptor-2 and fibroblast growth factor receptor-3in normal oral epithelium, epithelial dysplasias, and squamous cell carcinoma. *Oral Surg Oral Med Oral Pathol Oral Radiol Endod* 2002; 93: 573–9

[166] Waun K. Hong, and Reuben Lotan Increased Expression of Cytokeratins CK8 and CK19 Is Associated with Head and Neck Vol. 4.87/- 8'76. December 1995

[167] Werkmeister R, Brandt B, Joos V. Clinical relevance of erbB-1 and -2 oncogenes in oral carcinomas. *Oral Oncol* 2000; 36: 100–5

[168] Yanamoto S, Kawasaki G, Yoshitomi Di, Mizuno A. p53, mdm2 and p21 expression in oral squamous cell carcinomas: Relationship with clinicopathologic factors. *Oral SurgOral Med Oral Pathol Oral Radiol Endod* 2002; 94: 593–600

[169] YasuseiKudo, ShojiroKitajima, IkukoOgawa,MasaeKitagawa, Mutsumi Miyauchi, and Takashi Takata Small interfering RNA target ingo fSphasekinase Interacting protein 2 in hibits cell growth of oral cancer cells by in hibiting p27 degradation [*Mol Cancer Ther* 2005; 4(3): 471–6]

[170] Yasuyuki Nakamura, Manabu Futamura, Hirok Kamino, Koji Yoshida, Yusuke Nakamura, Hirofumi Arakawa: Identification of p53–46f as a super p53 with an enhanced ability to induce p53 dependent apoptosis: *J Cancer Sci*: 2006: 97: 7: 633–641

[171] Behrens J, Vakaet L, Frii R, *et al.* Loss of epithelial differentiation and gain of invasiveness correlates with tyrosine phosphorylation of the E-cadherin/ßcatenin complex in cells transformed with a temperature-sensitive V-SRC gene. *J Cell Biol* 1993; 120: 757–66

[172] Brachmann RK, Vidal M, Boeke JD. Dominant-negative mutations in yeast hit cancer hot spots. *Proc Natl Acad Sci USA* 1996; 93: 4091–5

[173] Brennan A, Boyle JO, Koch WM. Association between cigarette smoking and mutation of the p53 gene in squamous cell carcinoma of the head and neck. *N Engl J Med* 1995: 332: 712–17

[174] Datto MB, Li Y, Panus JF, *et al*. Transforming growth factor induces the cyclin-dependent kinase inhibitor p21 through a p53-independent mechanism. *Proc Natl Acad Sci USA* 1995; 92: 5545–9

[175] Dixon K, Kopras E (2004). "Genetic alterations and DNA repair in human carcinogenesis." *Semin Cancer Biol* 14 (6): 441–8

[176] Dowell SP, Hall PA. The p53 tumor suppressor gene and tumor prognosis is there a relationship? *Pathol* 1995 177: 221–224

[177] E M Rosen, S Fan and C Isaacs BRCA1 in hormonal carcinogenesis: basic and clinical research 533–548

[178] Fearon ER, Vogelstein B (1990). "A genetic model for colorectal tumorigenesis". *Cell* 61(5): 759–67

[179] Folkman J, Klagsbrun M. Angiogenic factors. Science 1987; 235: 442–9

[180] Hae-Ryun Kim, Russell Christensen, Noh-Hyun Park, Philip Sapp, Mo K. Kang and No-Hee Park Elevated Expression of hTERT Is Associated with Dysplastic Cell Transformation during Human Oral Carcinogenesis in Situ Clinical Cancer Research Vol. 7, 3079–3086, October 2001

[181] Harsh Mohan Essential pathology for dental students second edition 221–232

[182] H K Williams molecular pathogenesis of oral squamous carcinoma journal of clinical pathology 2000, 53: 165–172

[183] Hollstein M, Sidransky D, Volgelstein B, *et al*. p53 mutations in human cancers. *Science* 1991; 253: 49–53

[184] Dr. Irma B. Gimenez-Conti, DDS, PhD*, Thomas J. Slaga, PhD The hamster cheek pouch carcinogenesis model Journal of Cellular Biochemistry Volume 53, Issue S17F pp. 83–90

[185] Korineck V, Barker N, Morin PJ, *et al* Constitutive transcriptional activation by a ß-catenin–Tcf complex in APC–/– colon carcinoma. Science 1997; 275: 1784–7

[186] Kumar cotran Robbins pathologic basis of disease sixth edition 145–174

[187] Lee WH. Tumor suppressor genes—the hope [editorial]. *FASEB J* 1993; 7: 819

[188] Lianes P, Orlow I, Zhang ZF, *et al*. Altered patterns of mdm2 and TP53 expression in human bladder cancer. *J Natl Cancer Inst* 1994; 86: 1325–30

[189] Li-Kuo S, Vogelstein B, Kinzler KW. Association of the APC tumor suppressor protein with catenins. Science 1993; 262: 1734–7

[190] Liao PH, Chang YC, Huang MF, Tai KW, Chou MY. Mutation of p53 gene codon 63 in saliva as a molecular marker for oral squamous cell carcinoma

[191] Lucimari Bizari, Ana Elizabete Silva; Eloiza H. Tajara Gene amplification in carcinogenesis *Genet. Mol. Biol.* vol. 29 no. 1 São Paulo 2006

[192] Morin PJ, Sparks AB, Korinek V, *et al*. Activation of ß-catenin—Tcf signalling in colon cancer by mutations in ß-catenin or APC. Science 1997; 275: 1787–90

[193] Murry W. Hill and Christopher A. Squier Experimental aspects of oral carcinogenesis 492–501

[194] Ravi meharotra, Anurag gupta, mamta singh and Rahela Ibrahim application of cytology and molecular biology in diagnosing premalignant or malignant oral lesions.

[195] Sarasin A (2003). "An overview of the mechanisms of mutagenesis and carcinogenesis". *Mutat Res* 544 (2–3): 99–106. PMI

[196] Schottenfeld D, Beebe-Dimmer JL (2005). "Advances in cancer epidemiology: understanding causal mechanisms and the evidence Knudson AG (2001). "Two genetic hits (more or less) to cancer". *Nat Rev Cancer* 1(2): 157–62

[197] Shah, Johnson Batskin book of oral cancer 167–180

[198] Shiohara M, El-Deiry WS, Wada N, *et al.* Absence of WAF1 mutations in a variety of human malignancies. Blood 1994; 84: 3781–4

[199] Todd R, Chou MY, Matossian K, *et al.* Cellular sources of transforming growth factor-alpha in human oral cancer. *J Dent Res* 1990; 70: 917–23

[200] Wong DT, Weller PF, Galli SJ, *et al.* Human eosinophils express transforming growth factor alpha. *J Exp Med* 1990; 172: 673–81

[201] Yoshida T, Miyagawa K, Odagiri H, *et al.* Genomic sequence of hst, a transforming gene encoding a protein homologous to fibroblast growth factor and the int-2 encoded protein. *Proc Natl Acad Sci USA* 1987; 84: 7305–9

Websites:

[202] http://http://www.aetna.com/cpb/medical/data/300-399/0352.html

[203] http://http://ordonreserach.com/articles_cancer/Kobayashi/investigation1.html

[204] http://http://www.indmedica.com/journals.php?journalid=3&issueid=92&articleid=1268&action=article

[205] http://http://www.crcnetbase.com/doi/abs/10.1201/978020350.ch5

[206] http://http://www.ajol.info/index.php/eamj/article/viewfile/9038/1684

[207] http://http://www.path.cam.ac.uk/research/cancer–apoptosis.html

[208] http://http://www.crystalinks.com/cancer.html

[209] http://http://www.fmh.org/body.cfm?id=598

[210] http://http://www.fmh.org/body.cfm?id=598

[211] http://http://www.fmh.org/body.cfm?id=598

[212] http://http://telemedicine.org/warts/cutmanhpv.html

[213] http://http://www.answers.com/topic/carcinogenesis

[214] http://http://jpck.zju.edu.cn/jcyxjp/files/dao/04/MT/046M.pdf

[215] http://http://jpck.zju.edu.cn/jcyxjp/files/dao/04/MT/046M.pdf

[216] http://http://jpck.zju.edu.cn/jcyxjp/files/dao/04/MT/046M.pdf

[217] http://http://www.ncbi.nlm.nih.gov/pmc/articles/PMC1186964

[218] http://http://www.ncbi.nlm.nih.gov/pmc/articles/PMC1186964

[219] http://http://telemedicine.org/warts/cutmanhpv.html

[220] http://http://www.surgeryencyclopedia.com/S tWr/Tumor Marker Test Html

[221] http://http://www.surgeryencyclopedia.com/S tWr/Tumor Marker Test Html

[222] http://http://www.surgeryencyclopedia.com/S tWr/Tumor Marker Test Html

[223] http://http://www.surgeryencyclopedia.com/S tWr/Tumor Marker Test Html

[224] http://http://www.surgeryencyclopedia.com/S tWr/Tumor Marker Test Html

[225] http://http://www.surgeryencyclopedia.com/S tWr/Tumor Marker Test Html

[226] http://http://www.surgeryencyclopedia.com/S tWr/Tumor Marker Test Html

[227] http://http://www.surgeryencyclopedia.com/S tWr/Tumor Marker Test Html

[228] http://http://www.surgeryencyclopedia.com/S tWr/Tumor Marker Test Html

[229] http://http://akramania.byethost11.com/Robbins/60.html@printing=true.htm

[230] http://http://www.humanpath.com/spip.php?article14744

[231] http://http://www.ncbi.nlm.nih.gov/pubmed/11040937

[232] http://http://www.ncbi.nlm.nih.gov/pmc/article/PMC1186964

[233] http://http://www.ncbi.nlm.nih.gov/pmc/article/PMC1186964

[234] http://http://www.ncbi.nlm.nih.gov/pmc/article/PMC1186964

[235] http://http://www.ncbi.nlm.nih.gov/pmc/article/PMC1186964

[236] http://http://www.ncbi.nlm.nih.gov/pmc/article/PMC1186964

[237] http://http://www.ncbi.nlm.nih.gov/pmc/article/PMC1186964

[238] http://http://www.ncbi.nlm.nih.gov/pmc/article/PMC1186964

[239] http://http://www.ncbi.nlm.nih.gov/pmc/article/PMC1186964

[240] http://http://www.ncbi.nlm.nih.gov/pmc/article/PMC1186964

[241] http://http://www.bu.edu/histology/m/append02.htm

[242] http://http://jpck.zju.edu.cn/jcyxjp/files/dao/04/MT/046M.pdf

[243] http://http://jpck.zju.edu.cn/jcyxjp/files/dao/04/MT/046M.pdf

[244] http://http://jpck.zju.edu.cn/jcyxjp/files/dao/04/MT/046M.pdf

[245] http://http://www.ncbi.nlm.nih.gov/pubmed/16172191

[246] http://http://genes.atspace.org/28.4.html

[247] http://http://genes.atspace.org/28.4.html

[248] http://http://www.bu.edu/histology/m/append02.htm

[249] http://http://www.scielo.br/scielo.php?script=sci_arttext&pid=S1415-47572006000100001

[250] http://http://www.ncbi.nlm.nih.gov/pmc/article/PMC1186964

[251] http://http://www.ncbi.nlm.nih.gov/pmc/article/PMC1186964

[252] http://http://www.ncbi.nlm.nih.gov/pmc/article/PMC1186964

[253] http://http://www.ncbi.nlm.nih.gov/pmc/article/PMC1186964

[254] http://http://www.ncbi.nlm.nih.gov/pmc/article/PMC1186964

[255] http://http://www.ncbi.nlm.nih.gov/pmc/article/PMC1186964

[256] http://http://jpck.zju.edu.cn/jcyxjp/files/dao/04/MT/046M.pdf

[257] http://http://scielo.isciii.es/scielo.php?pid=S169844472004000500002&script=sci_arttext&tlng=en

[258] http://http://scielo.isciii.es/scielo.php?pid=S169844472004000500002&script=sci_arttext&tlng=en

[259] http://http://www.bioline.org.br/request?os08002

[260] http://http://telemedicine.org/warts/cutmanhpv.html

[261] http://http://bloodjournal.heamatologylibrary.org/content/99/4/1411.long

[262] http://http://bloodjournal.heamatologylibrary.org/content/99/4/1411.long

[263] http://http://bloodjournal.heamatologylibrary.org/content/99/4/1411.long

[264] http://http://www.ncbi.nlm.nih.gov/pubmed/12875969

[265] http://http://cancerres.aacrjournals.org/content/62/18/5295.full

[266] http://http://www.ncbi.nlm.nih.gov/pmc/article/PMC1186964

[267] http://http://www.ncbi.nlm.nih.gov/pmc/article/PMC1186964

[268] http://http://www.ncbi.nlm.nih.gov/pmc/article/PMC1186964

[269] http://http://www.ncbi.nlm.nih.gov/pmc/article/PMC1186964

[270] http://http://telemedicine.org/warts/cutmanhpv.html

[271] http://http://pmj.bmj.com/content/79/936/569.full

[272] http://http://www.path.cam.ac.uk/research/cancer-apoptosis.html

[273] http://http://mct.aacrjournals.org/content/4/3/47.long

[274] http://http://clincancerres.aacrjournals.org/content/11/6/2272.full

[275] http://http://portal.unitbv.ro.proxy/surf.aspx?dec=1&url=
uh44qwdELmSOOvOVPsqxBwhp8mCOLv5Z9vSVQtqVQMpeQMPelodwKTDpiva
=B6X

[276] http://http://www.biomedcentral.com/1471-2407/6/10

[277] http://http://www.ncbi.nlm.nih.gov/pubmed/1244439

[278] http://http://www.naturaltherapycenter.com/images/cancer/tumor20%markers.pdf

About the Author

Dr Manjul Tiwari, born on 13 Jan 1980 is currently working as Senior Lecturer, School of Dental Sciences, Sharda University, India from August 2009 after pursuing MDS from Kothiwal Dental College Moradabad in Oral Pathology and Microbiology.

He has over 40 publication on various fields of dentistry, genetics, forensic odontology in highly impact factor national and international journals ranging from Cancer therapies, Tumor Markers, gene therapy, child abuse, nanotechnology and forensic dentistry which has been published in Journal of Cancer Research and Therapeutics, Journal of Natural Science, Biology & Medicine, Journal of Oral and Maxillofacial Pathology, Indian Journal of Human Genetics, Guident, IDA Times, etc. He was invited Guest speaker in Hohhot, Inner Mongolia, China, Shanghai, Montreal (Canada) etc where he gave Oral Presentation on Tumor Immunology and Cancer Immunotherapy. Currently he is Reviewer in various journals like Journal of Oral and Maxillofacial Pathology.

He has collaborated Memorandum of Understanding with School of Bio-engineering, McGill University, Canada and State University of New York at Buffalo, (Buffalo University), USA.

He has attended various CDE's and workshops all over India. He has presented First and Second Prize winning Papers and Poster on "DNA Profiling or Fingerprinting" in XIX National & First International Conference of IAOMP, Radisson Temple Bay, Chennai, "Bite Marks: Role of Saliva in Human Genome" in XV National conference of IAOMP, Chennai, "Influence of inflammation on the polarization colors of collagen fibers in the wall of Odontogenic Keratocyst and their Clinico–Pathological Correlation" in VIII National Post Graduate Convention of IAOMP, Wardha and "Comparative analysis of primary intraosseous carcinoma & ameloblastic carcinoma: Case Reports with review of literature" in XVI National Conference of IAOMP, Khajuraho.

He is life member of several presitigious associations like Indian Dental Association (IDA), Indian Association of Oral and Maxillofacial Pathology (IAOMP), International Association of Oral Pathologist (IAOP) and International Association of General Dentistry (IAGD).

He has done various industrial training Program like on Nanotechnology, Biotechnology, Cancer Genetics etc, Advanced Fixed Prosthodontics, Smile design as well as he hold PG Diploma in Hospital Administration. His First International book on Alcohol, Tobacco and Oral Precancerous Disorders; ISBN: 978-87-92329-85-1; 2011; 1st Edition; 60 Pages; Rivers Publishers, Aalborg (Denmark) as well as National book on Modern Dictionary of Human anatomy from Deep & Deep Publications published 2010 and this book is his Second International publication by same Publishers.

Tumor Marker & Carcinogenesis

Tumor Markers
Carcinogenesis
Oncogenes
Cancer
Cancer suppressor genes
Apoptosis

Lightning Source UK Ltd.
Milton Keynes UK
UKOW07f1034171214

243278UK00001B/6/P